A Tale of Two Villages

A Tale of Two Villages

Coerced Modernization in the East European Countryside

Alina Mungiu-Pippidi

CEU PRESS

Central European University Press
Budapest–New York

First published in Romanian as "Secera si buldozerul"
by Polirom, in 2002

Published in 2010 by

Central European University Press

An imprint of the
Central European University Share Company
Nádor utca 11, H-1051 Budapest, Hungary
Tel: +36-1-327-3138 or 327-3000
Fax: +36-1-327-3183
E-mail: ceupress@ceu.hu
Website: www.ceupress.com

400 West 59th Street, New York NY 10019, USA
Tel: +1-212-547-6932
Fax: +1-646-557-2416
E-mail: mgreenwald@sorosny.org

Financial support for the translation was granted by the
Romanian Cultural Institute, Bucharest.

ISBN 978-963-9776-78-4 cloth

Library of Congress Cataloging-in-Publication Data

Mungiu, Alina.
[Secera si buldozerul. English]
A tale of two villages : coerced modernization in the East European countryside / Alina Mungiu-Pippidi.
p. cm.
Includes bibliographical references and index.
ISBN 978-9639776784 (hardbound)
1. Collective farms—Romania—History—20th century. 2. Farmers—Romania—History—20th century. 3. Romania—Rural conditions. 4. Romania—Social conditions—1945–1989. I. Title.

HD1492.R8M86167 2010
334'.68309498—dc22

2010008976

Printed in Hungary by
Akaprint Kft., Budapest

Table of Contents

List of Tables

Acknowledgements

This book is based on fieldwork carried out in Romania in 2001–2002 with a group of my students in the political anthropology class at the National School of Political Studies: Emanuel Răuţă, Victoria Timofte, Ion Naval, Stejărel Olaru, Ionuţ Ştefan, and Dana Ceauşescu. Interviews were also filmed, forming the basis of a BBC documentary I wrote and directed, "A Tale of Two Villages", broadcast by the BBC World Service in the summer of 2003. The original project was a collaboration between the Centre for Comparative Anthropology at the École des Hautes Études en Sciences Sociales (EHESS) and the Romanian Institute for Recent History. A grant from the Open Society Institute enabled the training of students and the fieldwork. The premature death of Gerard Althabe, originally my academic partner in this project, and the planned co-author of the Romanian and French versions, prevented his involvement beyond the stage of training the operators and reviewing my fieldwork. Some of the people I interviewed, including Elisabeta Rizea and Nel Preda, both survivors of the communist prisons, have also meanwhile passed away. The present version of this book is dedicated to all those who tried to help this story come to light. It is an update of the Romanian (Polirom) and French (L'Harmattan) versions, and has been translated by Angela Jianu from the Romanian and edited by the author.

CHAPTER 1

The Argument

Political change in rural societies has always appeared spectacular when observed from afar. Such frequent coups and aborted revolutions, grand reforms, and brutal assassinations are dramatic enough to attract journalists and scholars. On closer observation, however, it generates an almost unbearable feeling of monotony. Coups change only the person of the dictator; assassinations prove, sooner or later, to have been futile. There are many events and little evolution; change occurs, but development does not. Come back a century later and you will find, in the words of one Balkan historian, a "century of stagnation,"[1] regardless of what century you choose. Cities might occasionally push for reform; rural areas always tend to stagnation. And, as Samuel Huntington famously put it, he who rules the countryside rules the country.[2] Even regime change, despite managing to produce considerable suffering, does not modify the essential constraints under which every government will operate sooner or later. The future regime is contained in the past one, and, as in T. S. Eliot's *Burnt Norton*, all times seem unredeemable.

This book is a study of political change in rural Eastern Europe. The main question it attempts to answer is to what extent a program of social change imposed through vigorously sustained methods, including coercion, can endure once coercion ends. Is even the strongest state in the world able to change the essentials of a peasant society in a sustainable way? This study attempts to answer such wide-ranging, generic questions by focusing on a specific period and region where exceptional circumstances have rendered such behavior eminently observable: two East European villages under communism. These are two villages with exceptional—and very different histories—which in the end seem to have converged towards a similar outcome. As part of traditional "peasant societies" they have undergone, over a mere eighty-year period, colossal processes of political and social modernization meant, in the view of their initiators, to bridge the huge development gap separating their region from Western Europe. The size of this gap is suggested by a few comparative statistics: in 1930, Romania's per capita income was comparable to France's in 1789 or Britain's in 1648; in the early decades of the twentieth century, the sophistication and productivity of Romanian farming were at the level of French farming in the seventeenth century and of British farming in the eighteenth.[3] The political difficulties of this underdeveloped rural society were considerable, and the violent peasant revolt of 1907 sent shockwaves felt as far away as the chancelleries of Vienna and St Petersburg. What distinguished underdeveloped Romanian rural society from similar societies in other parts of the world was its relative proximity to the centers of European power. Its location at the periphery of Europe encouraged, more than in other regions, the persistent belief that this is a society with a European culture or at least a European vocation.

The study of rural Eastern Europe was seen as key to triggering its development a century ago. Seen from the West, it was a land problem; seen from the East, it was a peasant one, as David Mitranyi insightfully observed.[4] Since the fall of communism, it has received far less attention from scholars, with a few notable exceptions.[5] What is particularly missing is an exchange between political economists trying to explain the different performance of East European transition countries, political scientists working on voting behavior, policy scientists working on agriculture, and anthropologists interested in social change in rural postcommunist Europe. The older scholarship on the region thrived on precisely this bridging among disciplines, indispensable if one wishes to do justice to the importance of the rural factor in the economic and political development of Eastern Europe.[6]

This strong correlation between democracy and peasantry was established by the historian Barrington Moore Jr., who saw in the creation of non-repressive, commercial farmer agriculture the foundation of democratic development.[7] Traditionally, the remarkable resistance of peasant societies to change and progress has been explained by two different sets of causes. On the one hand, blame was laid on the peasant "culture." Peasants, as described by anthropologists in the twentieth century, are passive, collectivistic, envious, fatalistic, and distrustful creatures, clearly not the material democrats are made of.[8,9] Politicians held similar negative views on peasants; most modernizers, from the liberals to Vladimir Ilyich Lenin, looked upon peasants as the ultimate obstacle to social and economic progress. The second conception, based mostly on studies from the Third World, drew on the need to explain why peasants did not rebel against oppressive regimes. The conclusions of such studies were kinder to peasants, who were seen more as non-consenting victims than voluntary contributors to the conservative order of things. In a Latin American context especially, the rural upper class was identified by scholars as the main political opponent of democracy.[10] Oligarchs, usually landowners, were said to hold peasants captive, as their autonomy was too limited to allow the expression of their true political values. However, they resisted their captors through a variety of everyday resistance strategies.[11] Foot dragging, gossip, and stealing were no longer, in this view, expressions of the peasant character, but manifestations of protest when no other forms were available. The values of the peasants are, therefore, not conservative: peasants vote for their conservative landlords only because they are given no real choice.

In postcommunist Europe, too, the two sets of explanations have their champions. Clearly, the postcommunist agrarian social and class structures are different from both the traditional "junker" and "farmer" models which originally inspired this thesis. Large-scale mechanized, but collectivized, landholding, and the uncertain transformation of property relations after the fall of the old regime, have produced rural social structures which diverge considerably from Latin American models. The situation in postcommunist Europe also varies from country to country, with Poland, which was largely not collectivized, remaining rather an exception. Everywhere else decollectivization seems to have produced similar patterns: a return to family plots and subsistence farming, a "peasantization" of urbanites who became unemployed and resort to agriculture on their recuperated plots,[12] and a drastic fall in production as household consumption—not commerce—becomes the main use for crops. Peasants may have resisted collectivization strongly in the times of Stalin,[13] but after 1989 many resisted decollectivization and proved reluctant to accept markets. They long for subsidies and are rather opposed to the liberalization of land markets. A small percentage of farmers, owners of larger plots, who are more market-oriented, is gradually emerging, but nowhere is it higher than five per cent, Poland excepted. Land markets were slow to appear everywhere, due to logistic difficulties in restituting property, such as the absence of cadastral evidence. For the rest, differences prevail over similarities, with large state farms still important in Russia, medium-sized holdings much more numerous in the Baltics than in Central or South-Eastern Europe, and little to no property reform in Central Asia.[14]

Similarities can also be found when examining political behavior. Scholars working on voting behavior in postcommunist Europe have long pointed out that peasants tend, as a general rule, to vote for the wrong people. If a dictator is at hand, such as Serbia's Slobodan Milošević, they vote for him; if a candidate runs for an extra-constitutional mandate, such as Romania's Ion Iliescu,[15] they do not care; if a Communist Party is still around, it will more likely get their vote against any reformist party.[16] Peasants are also unlikely democrats, though old age and poverty, not some inherent peasant features, may be responsible for this.[17] In short, there is evidence which shows that peasants in postcommunist Europe behave in similar ways to peasants elsewhere, and that differences among countries may be due to the varying size of the peasant population, a legacy from pre-communist times. The opposite explanation, pointing to the informal institutional arrangements of the countryside as supportive of a pattern of abuse of peasants by predatory elites, was far less popular, though some anthropologists[18] and some political scientists[19] did encourage researchers to look beyond the "land" question. By and large, the reinvention of politics in the countryside has remained largely unexplored, and the little attention devoted to rural postcommunist Europe was absorbed by studies on land reform and agriculture.

One notable exception is a study by Kurtz and Barnes. The authors examine whether those states in which the agricultural sector is larger experience—all other things being equal—a more difficult time establishing open politics. The measure of agrarian dominance they use is the rural proportion of the population, as a rough indication of the size of the political base that might plausibly fall under the domination of the agrarian upper class (or its functional equivalent). They discovered that large rural sectors had a negative impact on efforts to liberalize politics, despite the peculiarities of the social organization of most communist agrarian systems. This result holds even when controlling for the level of socioeconomic development. "It is also important to notice that this effect is not due to the correlation between GDP and size of agricultural labor force, because the former is a poor predictor of democracy on its own. The causal mechanism here seems clear. Post-Communist agrarian elites have strong interests in retaining an authoritarian governance system, and they have inherited extensive organizational structures for quelling dissent and/or distributing benefits."[20] The only country deviating from this pattern is Poland.

If the rural factor is tested in models based on public opinion data, such as the World Values Survey pooled sample for East-Central Europe, residence in rural areas does remain a negative determinant of democratic orientation, controlling for age, education, and religion.[21] Residents of rural areas in the ten new EU member countries, Romania and Bulgaria included, show a lower appreciation of democracy than their urban counterparts. At the same time, it is crucial to note that they do not have a greater appetite for authoritarianism either: Rural residents do not endorse rule by strong leaders over rule by parliaments any more than urbanites do. They are significantly more traditionalistic and attend church services more often, they are more egalitarian, as shown by their disapproval of large income gaps, and they believe that the state should play a large role in society.

The meanings of concepts such as "peasant society" and "peasantry" are not similar in contemporary social science to the historical one granted by East European intellectuals of the twentieth century. Social science sees peasantry as a fundamental social category, only incidentally linked to a specific space or a particular historical period. Twentieth-century Romanian thinkers predominantly equated peasantry with Romanianness, anchored within a particular national specificity. There is a considerable difference between this vision of the village as the site of "Romanianness" as defined, for instance, by the Romanian interwar philosophers Lucian Blaga[22] and Mircea Vulcănescu, and the village as described by foreign travelers or seen by Romanian rulers, from the modernizing Hohenzollern monarchs to either agrarian or communist politicians. The former is an idealized, exceptional village, wherein external influences can at most shatter the illusion of an ahistoric, perfect balance; the latter is a real-life village, where impoverished people suffer and are deprived of basic resources, a place which can be saved only by a comprehensive, externally induced transformation. The former is incomparable, whereas the project aimed at turning the latter into a better living place is based precisely on its comparability to villages from developed countries and on the hope that development can be replicated through a set of prescribed steps. In its turn, anthropologists do not envisage the village merely as a self-sufficient community inhabited by people who earn their living by practising agriculture, but as a partial society with a partial culture: a rural society with a peasant culture within a larger society.[23] The word peasant "points to a human type. It required the city to bring it into existence. There were no peasants before the first cities. And those surviving primitive peoples who do not live in terms of the city are not peasants."[24] Kroeber's anthropological study of 1948 offered the earliest, brief and essentialized description of this society,[25] when he wrote that it differed from tribal society insofar as it lacked the latter's political autonomy, isolation, and self-sufficiency, and that it could be defined only in relation to the city, which provided a market for its farming produce. Peasant society, therefore, is one "half" of the whole, and cannot be understood in its own terms, without reference to the city. Peasant society is part of "a larger social unit (usually a nation) which is vertically and horizontally structured. The peasant component or this larger unit bears a symbiotic spatial–temporal relationship to the more complex component, which is formed by the upper classes of the pre-industrial urban center."[26] Peasants are therefore seen as a peripheral, but essential part of civilizations, producing the food that makes urban life possible, supporting (and subject to) the specialized classes of political and religious rulers and the other members of the educated elite. It is this elite which carries what Redfield called "The Great Tradition," giving continuity and substance to the sequences of advanced culture, and which lies in contradiction to the "Little Tradition"—which characterizes villagers themselves.

Most historical accounts which have contributed to a definition of peasantry focus on the medieval West European village, but that can be fairly compared to the South-Eastern European village in the modern age. We can, therefore, speak of peasants only if we locate them in a wider context, more specifically when "the cultivator is integrated in a society with a state—that is, when the cultivator becomes subject to the demands and sanctions of power-holders outside his social stratum."[27] Thus, the basic feature characterizing peasantry is its lack of power, its minimal control over circumstances governing its life.[28] The peasants' poverty and also—or especially—their lack of involvement in making decisions which affect their own lives thus form part of the anthropological definition of the peasantry.

In the case of underdeveloped societies such as Romania, with its limited number of urban centers—or with weak urban centers—the question remains whether the society as a whole could be considered rural. This distinction between a Little Tradition, embedded in the rural way of life, with practices which remain unchanged for centuries, and a Great Tradition, i.e. a national culture, created outside the countryside is difficult to make in the context of Balkan societies which were still overwhelmingly rural at the beginning of the twentieth century and had only a scarce elite. The twentieth-century Romanian thinkers who equated the Little Tradition with the national Romanian culture largely ignored the similarities between peasant cultures across nations, but they appreciated correctly the absence of a consistent "Great Tradition" in a country like theirs, which had acquired full independence from the Ottoman Empire only in 1877.

All these anthropological definitions might have retained a flavor of modernization and political development theories from the 1960s and 1970s. Meanwhile, the world has evolved on all counts: Poverty has decreased dramatically and the number of democracies is the highest ever encountered. Extensive modernization on all continents has certainly left fewer peasants in the world at large than there used to be. But peasants have not disappeared. In many countries they seem to be particularly resilient.

The present book examines the political condition of the peasant throughout this process of "managed" modernization, which communist regimes carried to its ultimate extreme. Managed modernization is not a natural process, such as the one which led to the emergence of West European civilization as we know it today. That was a process essentially based on technological and industrial development. By contrast, managed modernization is the attempt by domestic elites to replicate such historical developments with the declared aim of emulating and reaching the levels of Western development in their nations. In other words, we refer to an induced process of social and political change, which, in various forms, predated by fifty years the forced imposition of communism and continued afterwards. This process was invariably a reaction to the economic success of Western societies: Without the existence of this success model, it is certain that these elites would not have attempted to follow in the perceived footsteps of Western institutional development. They were guided by the hope that these would lead to economic and social development in their own society.

Political modernization is generally considered to be a two-tiered process. At one level, the aim is to create the nation's citizens by increasing literacy among peasants, thus replacing local dialects with the national language and fostering the emergence of a national consciousness. A second objective is the extension of full political rights—i.e. franchise—to the entire national body, starting with universal male suffrage. Social-economic and political modernization can be simultaneous or successive processes. As a rule, the West European model presupposes the primacy of economic processes over the political. Technological developments, industrialization, a high rate of urbanization, and generalized market relations preceded—and even required—political liberalization in the West. The English and French revolutions were determined largely by the need to bring in line the social and political rights of social groups which found themselves politically excluded, in spite of their relative autonomy and affluence. However, as Andrew Janos observed,[29] in the case of Romania and in underdeveloped countries generally speaking, this process works in reverse. Political modernization occurs prior to developments in other areas. Peasants are herded from the Little Tradition straight to the ballot box and into the arms of professional demagogues and the tussle of electoral propaganda. They are enrolled in adult evening classes to be educated by enthusiastic teams of city-dwellers mobilized by the modernizing regime, such as the teams of the sociologist Dimitrie Gusti, who taught them everything from basic hygiene to Romantic poetry. But economic and social development does not guarantee enough autonomy and affluence for the new citizens to become informed contributors to the political system and endorse its legitimacy. Instead, they are more often than not transformed into docile masses manipulated by unprincipled elites.

The processes of political modernization are fundamentally distorted under communism. Literacy programs are supported on a massive scale, as well as indoctrination into social and national values, but this is where the resemblance to modernization elsewhere stops. Communism imposes one single political choice, thereby annulling rights conferred by citizenship: those who do not accept this particular choice are liable to lose all other rights, including the right to life or freedom. Communist regimes also pursued social and economic development, the goal that interwar reformers had been following with methods ranging from state intervention (debt conversion or subsidies for agriculture for example) to enlisting volunteers to "enlighten" backward rural populations, but on a much wider scale and using more brutal methods. Their social intervention attempted to eliminate not only the class relations of pre-communist Romania, but also basic features of historical underdevelopment, which had much older roots. They targeted the lower urbanization, the absence of industry and, consequently, of a proletariat, as well as the comparative underdevelopment in most areas of life (when compared to the West), from economy to culture, all issues which had haunted Romania's pre-communist modernizers too. The attempt to use "wholesale methods" to sort out centuries-old deficiencies in a short time, in the belief that societies can compress developmental stages while largely following a universal model, characterized not only the communists, but also radical Romanian reformers from Mihail Kogălniceanu[30] to the Peasant Party. Their intervention in the key area of the growth of the farming population occurred very late, in the latter half of the nineteenth century, when the basic set-up of large-scale landholdings coupled with extensive farming—very different from West European practices, but typical of the East—had already been established. Although motivated essentially by the aim to create more equitable social arrangements and by the conviction that modernization, especially political modernization, was not possible otherwise, the "enlightened" intervention in the agrarian question led, via a series of reforms, to quasi-total failure.[31] The only success of these reforms was the ultimate emergence of a group—a mere minority among the peasantry initially, to become larger in the 1940s and 1950s—of economically emancipated peasants with the potential for becoming a class of rural farmers and entrepreneurs. For ideological reasons, it was precisely this class—the only group to resemble an autonomous middle class in a developed society—which was ultimately singled out for annihilation by the communist regime as the main obstacle in the way of communizing the Romanian village. The regime then proceeded to launch modernization proper, starting with investment programs (such as electrification), and continuing later with more ambitious and complex ones, such as the systematization of villages in the 1980s, which had a social as well as an administrative objective.

The contemporary traveler journeying from Tirana to Moscow and on to China is bound to recognize everywhere, despite some local specificities, the same typical landscape of the communist village, the joint product of social engineering and territorial planning. Such typical features are more pronounced, and the villages consequently more like each other, in areas where underdevelopment was more entrenched, irrespective of whether the inhabitants were, or still are, Orthodox Christian, Catholic, or Confucian. It is quite clear that the degree of transformation of a given society by communism depends on its prior stage of development. Using modernization as a pretext whenever it was a plausible option, communism was more invasive and distorted social organization more profoundly in underdeveloped than in more advanced societies. In Central Europe, especially in the Czech lands, where the degree of urbanization was comparable to that in Western Europe, the communist regime was constrained to build on pre-existing foundations and tolerate in the midst of its own project the remains of earlier history; in the rural areas and backward towns of Hungary, Romania, Bulgaria, Russia, or China, where stone had rarely been used for building, the traces of the past were totally effaced, to be replaced by brand-new villages, built on the tried-and-tested, unique Soviet model. In this process, the old manors of the former landowners were symbolically converted into head offices of collective farms (also called cooperatives or kolkhozes). In Romania, only the churches were left standing, but further east, even these were often demolished or converted.

The account presented in this volume uses the methods of political, sociological, and ethnographic anthropology to offer a suggestive picture of this unprecedented exercise in social engineering. My aim has been less to offer circumstantial explanations for situations specific to Romania than to present a model of coerced transformation and its consequences for individual and collective behavior. How can societies defend themselves against violence and abuse from the state? Is it possible to put an end to the cycle of dependency? How many of the lasting effects of coercive transformations were planned and how many resulted without the involvement of a structuring will? My chief fieldwork has been an ethnographic study of two Romanian communes, Nucşoara-Argeş, notorious for its inhabitants' resistance to the imposition of communism, and Scorniceşti-Olt, a commune famous for being the birthplace of Nicolae Ceauşescu, Romania's communist dictator from 1964 to 1989. These study sites were not selected for their fame; rather the reasons conducive to their fame also determined their selection for our research. Nucşoara, the headquarters of one of the fiercest groups of anti-communist partisans, was repressed with unparalleled violence. A mountain village, it remained uncollectivized. Scorniceşti, an impoverished village in the plains, was not only collectivized, but, as a great favor, systematized, and industrialized, a real guinea pig for the communist regime's successive agrarian policies. Together, the village in the mountains and the village in the plains encompassed a wide spectrum of the variation needed for the study, from individual property to collective property, and from maximum repression by the state to maximum investment. But despite the variation in conditions and in state intervention between the two villages, their politics at the time of their selection was absolutely similar. They both voted for the postcommunist party, they did not tolerate any political dissent in the village, and they were missing the former communist regime.

The field research was carried out in the summer and autumn of 2001. Unless otherwise indicated, interview excerpts in the book are taken from interviews conducted during that period by me and my team of students from the National School of Political and Administrative Studies in Bucharest. However, in order to be able to generalize my findings I also resorted to a larger survey, representative of the entire Romanian population, of the rural population, and of the population of the Argeş and Olt counties (and of every other county) separately. This mega-survey was conducted among 37,474 respondents by professional pollster CURS (Romanian Centre for Urban and Regional Sociology) in October 2000 (Appendix 1) on the basis of a questionnaire I designed. I also quote occasionally from other national representative polls that I have designed and carried out during my work as National Coordinator for the UNDP early warning system program.

This book is structured as follows: Chapter two is an overview of the basic features of the two locations, from geography and social structure to their exceptional history. Further on, I analyze one by one four strategies of domination identified as essential in the creation of a dependent peasantry: the manipulation of property (chapter three), of social conflict (chapter four), of access to collective resources (chapter five), and of lifestyle (chapter six). In chapter seven I examine the impact of these communist strategies on the Romanian village, and especially on its social structure and political behavior, as well as the resulting model, which I have called a model of neo-dependency. In the last, eighth chapter, I survey the agrarian policies of Romania in transition before and after accession to the European Union in order to assess the future of the village and of the peasantry.

Social engineering on a gigantic scale, as practised under communism, is a rare occurrence, and optimists might think that such exercises belong to the past. This does not mean that mechanisms comparable to those I describe in this book cannot be reproduced in another circumstance where a predatory elite gains access to similar resources, even though it may be in a more restricted space, and with fewer lasting effects. The causes of evil are, of course, the concern of theologians, not of social scientists. But the latter nevertheless have learned from fascist and communist experiences, and particularly from the Holocaust, that evil happens, and that we must have some theory to explain its causes and effects. Societies can survive violence, but the cost of survival is often underestimated by victims and observers alike. Besides its modest contribution to social science, this book is also meant to bring to light a tragic and original collective experience which is worth the public knowing about—even if its survivors themselves would prefer to forget it.

Notes

[1] Palairet, *The Balkan Economies.*
[2] Huntington, *Political Order in Changing Societies*, 292.
[3] See Clark, *The Conditions of Economic Progress*, 41, 83.
[4] Mitranyi, *The Land and the Peasant.*
[5] The works of Verdery, Kideckel, Chirot, and Klingman are notable in this respect.
[6] For the older tradition see Mitranyi, *The Land and the Peasant*, and idem, *Marx Against the Peasant*; Stahl, *Satele devălmaşe*; Roberts, *Romania.*
[7] Moore, *Social Origins of Dictatorship and Democracy.*
[8] See Redfield, *The Primitive World*, and idem, *Peasant Society and Culture.*
[9] See Foster, *Peasant Society and the Image of Limited Good*, and idem, *Peasant Society.*
[10] O'Donnell, *State and Alliances.*
[11] Scott, *Weapons of the Weak.*
[12] Leonard and Deema, *Post-Socialist Peasant?*
[13] Fitzpatrick, *Stalin's Peasants.*
[14] Wegren, *Land Reform.*
[15] Ion Iliescu enjoyed two terms as the first postcommunist president of Romania between 1989 and 1996; he was re-elected for yet another mandate between 2000 and 2004.
[16] See Mungiu, "Intellectuals as Political Actors," Wegren, *Land Reform*, and Gordy, *The Culture of Power in Serbia.*
[17] Rose et al., *Democracy and its Alternatives.*
[18] Humphrey, *The Unmaking of Soviet Life.*
[19] Jowitt, *Social Change.*
[20] Kurtz and Barnes, "The Political Foundations," 545.
[21] Mungiu-Pippidi, *Politica după comunism.*
[22] Lucian Blaga (1895–1961), Romanian poet and philosopher. In his *Trilogia culturii* he presents rural Romania as the shelter of the true "ethnic being." He makes a distinction between two tiers of culture. On one side were the towns, the sites of political power and of interaction with foreigners. On the other was the alternative village culture, perpetuated in unbroken continuity with the past and therefore the depository of true Romanianness. Blaga was persecuted in Stalinist times, but during the national communist regime of Nicolae Ceauşescu his work was largely rehabilitated.

In the interwar period, this school of thought, which constructed an imaginary village as a site of the national ethnic spirit, coexisted with a powerful intellectual movement which had emerged around the "sociological school" of Dimitrie Gusti (1880–1955). For Gusti, the village was a socioeconomic field which, with its modes of domination and exploitation, was the product of history. For this school, research into the rural world was a science steeped in a comparative–interdisciplinary approach the practical objective of which was to modernize villages. Henri H. Stahl (1901–1991) was the pre-eminent representative of this modernizing project.

23 Kroeber, *Anthropology: Race, Language, Culture, Psychology, Prehistory*, 284.

24 Redfield, *The Primitive World*, 31.

25 Kroeber, quoted in Foster et al., *Peasant Society: A Reader*, 2.

26 Foster, "What is Folk Culture?," 163.

27 Wolf, *Peasants*, 11.

28 Foster et al., *Peasant Society*, 8.

29 Janos, "Modernization and Decay in Historical Perspective."

30 Mihail Kogălniceanu (1817–1891) was a man of letters and major statesman, a promoter of the first land reform in Romania.

31 Roberts, *Romania*.

CHAPTER 2

Two Villages

Countryside landscape

Scorniceşti is a village dispersed across the plains between the provinces of Oltenia and Muntenia. The nearest European motorway, joining Piteşti and Craiova, is only fifteen minutes away, but it circles the village from afar; there is no railway station. As the local authorities are keen to point out, the advantage is that Gypsies[1] can reach the village only by horse-drawn carts, and then they are likely to be intercepted by the police and turned away. But there is a downside too: The village is cut off from the wider world. Turkish armies came near on a punitive raid following the 1821 uprising led by Tudor Vladimirescu.[2] They set fire to a handsome church in Constantineşti, the ruins of which still stand defiantly. General Mackensen occupied Oltenia for almost a year in 1917–1918,[3] and was appalled by the poverty of the inhabitants. The sympathetic German general ordered the conquered barbarian territory to be civilized and, to this end, had public lavatories erected throughout the region. Made of metal sheeting cut to identical size, they were meant to stop the locals from relieving themselves in ditches and on public roads. Twenty-four hours after the withdrawal of the German army, these structures were torn apart, and locals each took a piece to proudly patch their rooftops.

Scorniceşti is a fiction: Nothing stands scrutiny there, not even the name. The old village, entirely demolished during the "systematization" of the 1970s and 1980s, used to be called Tătărăi. During the administrative reorganization, Ceauşescu, who was born there, ordered it to be turned into a mega-commune by merging four adjacent ones and making "Scorniceşti" the capital: It got a brand-new community centre to mark the occasion. Scorniceşti was thus created out of a merger of the communes Negreni, Mărgineni, Mogoşeşti, and Scorniceşti. In 1989, during the final countdown of the communist regime and of Ceauşescu, a son of the village, it was proclaimed a town. Today, it has 12,787 inhabitants, 4,798 of whom live in the centre—in Scorniceşti proper—and the rest in the adjoining villages. It is Romania's most dispersed village, with a surface area of 139 square kilometers, and encompasses fourteen villages. Of these, Negreni is the oldest commune and is highly self-conscious: When they were allocated to Scorniceşti, the locals felt downgraded. Rightly so: Whereas the secondary communes declined for lack of investment, Scorniceşti itself prospered and became a small town. The price was high: People lost their homes and their gardens and were relocated into apartment blocks. However, today the town is still productive, whereas the adjoining villages, living solely off agriculture, are declining. Mindful of the independence-seeking ambitions of the inhabitants of Negreni, the mayor of Scorniceşti set them up with a local referendum, in which over ninety per cent voted for separation. However, as today villages profit from funds redistributed from the town, once they got their dreamt-of emancipation they were destitute. As the mayor of Scorniceşti points out though, winning or losing from this separation is irrelevant. It was simply the time for those people to master their own lives. The temporary union with Scorniceşti dating from socialist times was to cease, and it was every man for himself in the new capitalist order.

The commune of Nucşoara lies to the north of the county of Argeş, seventy kilometers away from the county town and forty kilometers from Curtea de Argeş. With a surface area of 43,709 hectares, it comprises the villages of Nucşoara—as its chief commune—Gruiu, Sboghiţeşti, and Slatina, with a total of 1,800 inhabitants. Nucşoara is an ancient village, or at least its inhabitants think so. They claim that the village was mentioned in a document issued by the prince of Wallachia on 25 June 1547 to endorse the title deeds of a property. It is not clear whether the document refers to this particular Nucşoara, or to another one in a county further east, as inhabitants there could also swear that the document refers to them. It is true that both regions, Argeşul de Sus (Upper Argeş) and Haţeg county, are recorded as having been inhabited from an earlier date than other areas, but today Nucşoara-Argeş is famous for more recent events. It was the last bastion to fall to communism: A handful of anti-communist partisans resisted, with the passive support of villagers, for over ten years in the adjoining mountains.[4] No signs of earlier times are to be found in the present village. However, not far from there, at Brădet, beyond the hill which the partisan commanders, Toma Arnăuţoiu and Gheorghe Arsenescu, chose as their hiding place, there is a fifteenth-century monastery dating from the times of Voivod Mircea the Old. At Curtea de Argeş, the nearest town to the south, stands another of Wallachia's oldest churches, St Nicholas. On its walls someone inscribed in 1352 the death of the state-founding prince, Basarab.

The region lies on the southern slope of the Carpathians, where hills descend slowly from the Făgăraş mountains. The river Doamnei runs through it. The land is beautiful, though savage. Few people come to admire it: Access is possible only along regional roads of abysmally poor quality. In 1997, the first anti-communist government after the 1989 Revolution decided to honor the martyred village and the Ministry of Public Works asphalted the three kilometers of the road going uphill to Nucşoara, from Sboghiţeşti to the village hall. Further away, towards Brădet, the native village of Emil Constantinescu—Romania's president in 1997—the road breaks down completely, while the journey from Domneşti to Sboghiţeşti, some fifteen kilometers, is possible only by tractor. In spite of its bright patch of asphalt, Nucşoara remains isolated, as it has always been. The Curtea de Argeş-Câmpulung motorway which passes through Domneşti is the nearest link to civilization.

Landscape matters. The Romanian poet and philosopher Lucian Blaga constructed a theory that a people's psychological profile derives from the geographical landscape it lives in. Gradually, he argued, that landscape will exert its enduring influence upon people's "nature," another concept which features prominently in Romanian self-reflective literature, the best represented written genre in the country's history. This characteristic landscape is allegedly composed of a smooth succession of hills and valleys, a gentle, balanced landscape which encourages a "culture of survival," the name given by Blaga to the Little Tradition, which he considered more resilient than the Great Tradition *sub specie aeternitatis*. In Blaga's own words: "A minor culture, born of improvisation and spontaneity, and endowed with little taste for eternity, stands a better chance, in its stillness, of lasting for thousands of years [...] whereas a major culture, stemming from an ambition to conquer space and time is, precisely because of its dynamism, far more prone to disasters and decline."[5]

Although not exactly conveying a sense of "taste for eternity," in Nucşoara and Scorniceşti alike the landscape is indeed peaceful and serene. And although one village lies at the foot of the Făgăraş, the highest peak in the Carpathians, and the other in the lowlands of the river Olt, there are times in the day when the two become very much alike. From the top-of-the-hill cemetery where all members of the Ceauşescu family are buried—excepting the dictator Nicolae and his wife Elena, who were rushed off to a military cemetery after their execution—the tall blocks built under communism on the site of the old Scorniceşti cannot be seen, hidden as they are behind tall acacias and hillocks. The only visible bits of scenery are the smooth slopes merging into the lush lowland below. Beyond the edges of the village, the roads are lined by scattered acacias: as far as one can see, the area has been deforested. Leaving Scorniceşti to search for firewood is a real adventure. By comparison, in Nucşoara, the forests which for ten years sheltered partisans hiding from the communist regime still extend as far as the common pastures. These are still communally owned in the same way they were in England before the eighteenth century, and encircle the villages like oases of grazing land bordered by forest. It was enough for Verona Jubleanu to step into her garden and look up to spot her brother stepping out of the forest shadow. This was when he needed to signal that provisions had run out. She would then go up the pathway to take fresh supplies for him. Similarly, Vică Berevoianu would look up from his garden and, if there was movement on the edge of the forest, he knew that the "boys"—the term of endearment for the partisans—had run out of cartridges. He would then look for a new supply in the earthenware pot where his mother collected the ashes for leach-making and he took them uphill. This pasture was also the way for Marinica Chircă, a woman who liaised with the partisans, whom the Securitate[6] called "bandits." She would shuffle quietly after sunset past the garden wall of a man who was later sentenced to long years in prison simply for the crime of failing to notice and denounce her. The communists installed a weather station above the common from which they were able to survey everything that moved, but the villagers themselves soon learned how to monitor every movement of the intruders. The landscape, hazardous and covered by forest, proved, for more than nine years, the ideal shelter for those on the run.

The rivers and cemeteries may look like an ante-historic pastoral space, but only from selected angles. Wild lilies grow randomly on the graves of those executed by the communist regime in Nucşoara as well as on those of Ceauşescu's close relatives in Scorniceşti, who had for thirty years ruled over the village with an iron fist. From the top of both cemeteries, as far as one can see, there are only valleys and hills, and even the difference in temperature between the village of the cooler mountain area and that of the warmer plains is blurred there. But history has left a visible legacy: The inscriptions on the graves of the Arnăuţoiu brothers, the chiefs of the "bandits" of Nucşoara, are new. It is quite obvious why they could only be added after 1990, although the brothers were shot dead in 1959. Nicolae Ceauşescu's sister, Elena Bărbulescu, died in 2001, while I was conducting research in the village. She had practically run the commune, in fact the entire region, jointly with her brothers, husband, and lover. The fresh grave was covered in funeral wreaths ordered from the town Slatina, and from even more distant towns. Eleven years after the fall of communism, one of the ribbons carried an enigmatic tribute from the "Kolkhoz Scorniceşti."

Should one gaze over the lakes at the break of day in mid-July, for instance, when in Nucşoara it is still very fresh, while in Scorniceşti it starts to feel like scorching summer, the landscape feels equally peaceful, in spite of its variety from one village to another. In Nucşoara, the lake is surrounded by widely spaced houses and gardens, and the landscape is terraced: From a hill, the renovated house of a villager, with its balconies overlooking the lake, might as well be somewhere in Switzerland. Tall reeds hide the fact that nobody bothers to clean the lake, and flocks of wildfowl swarm there undisturbed. In the other village, in Scorniceşti, the landscape is arid, the water is clearer and sparkles in the sun, and the golden pond lies enclosed within margins carefully trimmed by man's hand, like an oyster within its shell. Here there are birds, too, including seagulls, although the sea is far away, 600 kilometers. The place is dotted with frail-looking saplings, planted by an unskilled landscape gardener. All traces of the ancient forest, cut down in communist times to free land for farming, have disappeared. From the edge of the lake, in any direction one looks—apart from Scorniceşti itself, hidden behind a hill—there are no trees, only acacias and lilac bushes growing close to garden hedges. However, though difficult to compare, both lakes bear the marked traces of human intervention under communism. In Nucşoara, the fences of individual properties are cut across by a circular fence, creating a path along the edge of the lake, with moorings for the fishermen. The local socialist councilors had to return lands around the lake in the property redistribution program which started after 1990, but they decided to expropriate a band of land for communal use, so that fishermen should not have to ask for the owners' permission every time they wanted to fish. One owner of land on the lake shore, Nel Preda, a former political prisoner, tried to fight the decision in court for as long as he could, rather than give the former communists this final satisfaction. But he died of a heart condition in the winter of 2002 and now harmless fishermen, with no idea of local history, sit on his garden fence and try their luck. But there is very little fish, in either village. In Scorniceşti, the lake's patron keeps replenishing the fish stock, but it literally cooks in the water on hot summer days when temperatures rise above 42°C. In addition, the peasants, who disapprove of strangers coming and buying land in the area, keep throwing insecticide and other chemicals into his lake whenever they have a chance. Not very often: Nicu, the patron, now pays bodyguards, endowed with mobile phones, electric stun batons, and all sorts of safety equipment to keep people from throwing rubbish into the lake and killing the fish. But it is virtually impossible to eat fish from either lake. In Nucşoara, if you insist, you might get some trout caught upstream in the river Doamnei. Around Scorniceşti, the nearest rivers, such as the Olt, are seriously polluted, so the only fish available is frozen tuna supplied from Slatina in refrigerator trucks. On Nicu's terrace, meant to become a big international restaurant when the fame of Scorniceşti will finally reach the world, one can safely eat only American chicken thighs, from international aid and resold in Romanian shops. At daybreak, the peace of the place is shaken by shooting parties. On organized hunting trips Italians shoot the game under the contemptuous gaze of the waiters, for they are extremely bad shots. Armed with a special permit, the tourists go shooting wild boar, chased by beaters straight within their shooting range. But the Italians do not really come here for the hunt. The real trophies are the girls, clumsy waitresses in the daytime, gauche escorts at night. But they are young and fresh. The boss does not want them to appear on the publicity brochure for the inn, a draft of which is being created on one of the terrace tables, with my unwitting participation. The cover will feature the "Brooklyn Bridge," my nickname for the absurd Ceauşescu period chain suspension bridge which links two corners of the lake, from Nicu's terrace to the former hostel of the kolkhoz, now under refurbishment. The bridge sparkles with lights at night. Nicu does not want the girls to be on display: This would arouse suspicion, especially in Italy, and the visitors would not want people to know that their package holiday included a lake without fish, half-tame farm game, and amateur prostitutes. Particularly since his tourists have families back home.

The landscape bears the marks of the regimes of yesterday and that of today. In Nucşoara, by the bank of the lake, in a reserved area on the motorway, stands a monument built by those who paved the road. A plaque awkwardly tied to a pile of boulders claims (in huge letters) that the monument was erected in 1998 by the Ministry of Labor and Territorial Administration to honor (in lower case) the heroic resistance of Nucşoara's villagers against communism. The locals are grateful for the new asphalt and many regret that they did not take further advantage of the anti-communist regime which lasted from 1996 to 2000. But they look puzzled when queried about their fight against communism. "Well, they found me guilty in a communist court, but to say I was anti-communist, that's an altogether different matter," says Vasile, the father of Nucşoara's mayor, who, as a young boy, was minding the sheep when the partisans came to get cheese from the shepherds. They thought communism had only months to live—as Radio London and Radio Paris kept announcing at the time—so they scrupulously kept lists of everything they took, which they regarded as temporary loans, to be returned after the fall of the regime. When they were caught, all their registers were found by the Securitate and this is how the boy Vasile, who could barely sign his own name, became an anti-communist militant. He looks back stoically: He was not the only one in this situation. The prescribed prison term for political offences, he explains hilariously, could be summed up anecdotally as "Five cigarettes, ten years." This was the sentence passed on a villager who chanced on the Arnăuţoiu brothers in the mountains and gave them five cigarettes. Those cigarettes sent him to the gulag. Today, Vasile is upset mainly because the other former political prisoners are accusing his son of betraying the historic National Peasant Party (NPP) and joining the Social Democratic Party (PSD), the successor party to the communists. His accusers claim that this was the price of his mayoral position. Vasile, however, says that, like many others who served prison terms, he was no more of an anti-communist than his son is a communist now. He is simply a man who wants to put things right in the village, and the rest know that you cannot be a mayor unless the former communists agree. Peasants seek only their own advantage and listen to no-one and have lost all good old-fashioned decency, Vasile says ruefully. The ditches alongside the new road are blocked with leaves. When the mayor calls upon the people to come out and clean them up, they shout back that he is a communist and why doesn't he just go away. The mayor's office does not have a budget to hire workers, and why should it do it? Every man should be able to clean up the front of their gate, Vasile says crossly. They've got plenty of time on their hands, they spend their days fishing. The villagers merely shrug their shoulders. Many think the mayor is a man of limited abilities, others say that this was precisely why they voted for him: The peasant wants a leader he can easily manipulate. However, everyone agrees philosophically that it is not the mayor's fault. It is the "people's."

Ceauşescu does not have a monument in Scorniceşti, although many think he would deserve one. But the town today is run by pragmatic people who would do anything to efface all traces of the times when this place was an object of fear and envy. They do not think that just because Ceauşescu happened to be born here they are any more to blame than the rest of the Romanians. "Those who blame us have calloused hands from so much applauding at party meetings," says the mayor Tiberiu Mateescu, and he is probably right. However, some of his associates defend enthusiastically the memory of the late dictator, even though they have aligned themselves to the new times. One of them is the local cultural animator, Mr Luțu, the director of the local Culture House both under the old regime and today. He believes that Ceauşescu has earned his place in Romanian history. "Ceauşescu's time cannot be erased from history, just as you cannot erase Scorniceşti's history," he says with some grandeur. "Good or bad, Ceauşescu's time is worth recalling," he says. It is the historians' task, he believes, to write both the history of Scorniceşti—with himself, Mr Luțu, alias Nelă, as the local worthy—and the history of Romania, by giving Ceauşescu his due. The old house of Ceauşescu's parents, refurbished in the 1980s as an example of Oltenian folk culture, now stands empty and locked, with only a few black-and-white photographs pinned to the walls. In the pictures, the Ceauşescus are surrounded by people who today would not dare admit that they had ever known them. The nephew Emil, the son of Elena (Lina) Bărbulescu, has no money left to maintain the property, so the electricity company cut the house off. Occasionally, though not too often, foreign cars stop under the acacias and tourists take pictures of the house over the fence. People were afraid of Lina up to the last day of her life. Today, they are much more relaxed, having seen her off to the cemetery up the hill and having spent more than their quarterly electricity bill on funeral wreaths for her. All public institutions sent wreaths, starting with the high school—her creation—the only agro-industrial vocational high school in the area. Everybody was relieved, but they sent her off with great pomp.

Is the village world as isolated and reserved vis-à-vis strangers as is generally believed? Nowadays, villagers, even in Eastern Europe, have television and radio. Politicians canvas in villages during electoral campaigns. People in Nucşoara are very proud that both King Michael and President Emil Constantinescu came to see Elisabeta Rizea, one surviving heroine of the resistance, although the villagers do not like her. In Scorniceşti they were pleased that the head of the Liberal Party campaigned there, although nobody voted for him. Otherwise, however, in terms of perceptions, distances are considerable, especially in the mountains. If you enquire about somebody in Nucşoara, for instance, you may be told that they are not locals, that they are from Sboghițeşti, but the village of Sboghițeşti is only two kilometers downhill, and is part of the same commune. Many villagers have come to own cars, national Dacias, but there are still no regular bus services to the nearest towns, Câmpulung, which is close to Nucşoara, and Slatina, close to Scorniceşti.

All peasants, in Scorniceşti as well as in Nucşoara, complain, saying that life was much better in Ceauşescu's time. As a mountain village, Nucşoara was never collectivized: There was an attempt at a joint farming venture, but the associates did not do well and had to sell the sheep. All around them, even in Domneşti, the larger commune down the valley, cooperatives were set up, so people lost their cattle and ended up depending on the people of Nucşoara for food. The inhabitants of Nucşoara did very well out of this, as traders from as far as Curtea de Argeş would come to buy cheese from them. The same economic policies which had starved towns in the 1980s led to prosperity in this uncollectivized region. In Scorniceşti supplies were exceptionally good because of the village's privileged status. Bread was rationed in Constantineşti, explains mother Stancia, but in Scorniceşti bakers had plenty of bread rolls to sell. "We used to go there and get bread rolls, and they kept coming," she says with a shrewd smile. "We lived nearby, so why not take advantage? I always had something to take with me to Bucharest when I went to the doctor's for my rheumatism. Nurses there kept moaning that there was no food to buy and everything was so expensive. Here, we had everything: chicken, innards. We could manage." Mother Stancia is the wife of one of the few political prisoners from Scorniceşti, who served time because he opposed collectivization. There were not many of them, just a few reckless youths who planned to set up an organization called the "Olt River Voice" to campaign against collectivization, and some landowners who were later arrested simply because they owned land. Each member of the projected organization had to enlist four others, and each of those a further four, and so on. They had barely reached five members when the voice of the river spoke up prematurely and they got busted. They received a year each, at the end of which their sentence was commuted. After this, the villagers joined the kolkhoz *en bloc*. The communists wanted mother Stancia, as a "bandit's wife," to join first, to set a good example for the rest. She had begged them to let her stay out. But in the end the Stancia husband and wife were reconciled to their fate: He became a mechanic at the cooperative, did well for himself and worked for Lică Bărbulescu, Lina's husband. Today, their daughter is married to the owner of a distillery. The young couple managed to have a large portion of land returned to them after 1989 and bought a tractor, all of which makes them the object of envy in the village. They had another storey added to the house, and bought expensive sofas in town, which are still wrapped in the plastic sheets in which they were delivered, and so they will remain. The peasants, in any case, will not sit for fear of soiling them. In many villages, people still used to have their "quality" house kept as an annex, with a large and well-furnished room ready for guests, while the owners slept above the barn in the yard. Nowadays, they have evolved: They build new houses with attic bedrooms and terraces, and they buy imitation leather sofas. But still they do not dare sit on them.

Both villages were electrified in the 1970s, but today they are declining fast. Only the urban centre of Scorniceşti is still a focal point for the locals, but most teenagers dream of going to university and escaping their peasant condition. They all want to leave, but it remains rather a dream. However, the surrounding villages, both here and in Nucşoara, have been gradually emptied of people, an irreversible process which could not be stopped even by the inverse migration caused by the industrial recession of the 1990s. In the vicinity of Nucşoara there is no high school, and the children who attend school in town get a taste of the good life and do not return. The population is ageing. There are only seven or eight weddings annually in Nucşoara. The village priest complains that his flock is diminishing daily. "I've got 554 souls all in all, 419 souls in Nucşoara and 135 in Gruiu. There are 175 families, and 75 widowed people in Nucşoara, and 31 families and 31 widowed individuals in Gruiu. The parish is declining at an alarming rate. Because they are mostly impoverished people, we no longer charge funeral taxes." The young are fleeing poverty and the rural lifestyle. In the communes adjoining Scorniceşti, one in three houses is boarded up. The Spinei area and the northern parts of the county are particularly impoverished. All the young people have left. Some of the elderly are helpless: They have children in Bucharest who visit them only once every three months.

The only ones to stay, because they cannot find work elsewhere, are the so-called *rudar* Gypsies from Nucşoara. The descendants of the boyars' Gypsy slaves from Argeşul de Sus (Upper Argeş), they received very small allotments at the end of the war, supplemented by bits of land confiscated from former political prisoners, and today live mostly off "foraging." The name *rudari*, they explain, denotes people who do woodwork, making things such as sheds, gateposts, and utensil handles. These carpenter Gypsies claim to be descended from "the Dacians or something like that, 'cause we speak the same language. Only in pronunciation we have an accent. We don't think of ourselves as Gypsies, because we don't speak the Gypsy language, but there's a lot of racial hatred around." It is enough, however, that others think of them as Gypsies. When they go on business to the village hall, the secretary barely looks at them. They live in Sboghiţeşti and Slatina, where they huddle in miserable, derelict houses along the motorway. Their women pick wild mushrooms for the national Forestry Authority, which usually pays them less than half a euro for a whole day's pickings.

The mayors of both villages nurture ambitions as sponsors of the local tourist industry. Nicu's commercial enterprise is supported by the mayor, who wants to add a radio station on the site. On Saturday evenings, many people from outside Scorniceşti come here to dance in a natural cauldron-shaped spot by the lake, under the sparkling bridge, where Ceauşescu himself planned the building of a lido many years ago. The mayor's office already has EU PHARE-sponsored programs[7] under way, and they will undoubtedly find ways of attracting tourists, as they have attracted the Italian hunters, for instance. But they do not intend to use Ceauşescu's image for the publicity, in the way that the Ministry of Tourism uses Dracula: It would be undignified, they say.

The mayor's office in Nucşoara does not have a PHARE program in place, and nobody seems competent enough to fill in one of those complex application forms with many rubrics, seemingly designed to exclude the semi-literate farmers of the underdeveloped, rural areas for which those programs were in fact intended. The land itself costs almost nothing—an acre on the banks of the lake is 200 euros. People have heard of the availability of European funds for local tourism and of programs which encourage peasants to turn their homes into B&Bs, but they are skeptical, largely because of the poor condition of the road. In Argeşul de Sus many people from Bucharest bought land and had villas built. Even around Nucşoara, one can see the occasional amateur fisherman with a Bucharest license plate: They come for the weekend and are stunned by the beauty of the place. When I was there, a cable TV company from Curtea de Argeş sent some adventurous technicians up the mountain path in a van, and these forerunners of civilization brought cable to many households. Now that they no longer pay funeral taxes, people can afford to pay three euros per month, the cost of cable subscription. After all, as Nel Preda told us shortly before he died, on the evening when—to his and his wife's great joy—I helped him select his cable TV channels: It was not so much for the villagers as for their children. If the parents did not have cable TV, the chances of the young generation ever returning for a visit from town would diminish dramatically. Now, with Eurosport, MTV, commercial breaks, and football matches, the differences between countryside and city have become blurred. The priest himself had had an enormous satellite dish installed in his garden. The numerous channels were intended, he said meekly, "for the boy, to help him learn a foreign language."

The behavior of these impoverished peasants, whose expenditure on toiletries and personal hygiene is at third-world levels, is not exceptional. All over Romania, television replaces hot running water, the job, and any forms of entertainment. A national poll indicates that sixty per cent of the unemployed have a cable television subscription, a figure which does not include those who have simply adapted their TV sets to capture their neighbors' subscribed channels.[8] Cable coverage in urban areas stands at over eighty per cent, and rural areas have started to catch up in recent years. Most channels do not have TV shows targeted specifically at the rural population. Only state television airs a couple of folklore music shows and a debate on farming entitled *Country Life*, which was inherited from the communist period with little change of format. But peasants go to bed early in any case, so they mostly watch the evening news and the soaps, from Latin American series to *Sunset Boulevard*. Advertising agencies are not concerned with absolute audience numbers, and consider only audiences with higher purchasing power anyway. For manufacturers of detergents, feminine hygiene products, or expensive drinks, not to mention cars and electrical household appliances, the purchasing power of rural populations is zero. From their viewpoint, a TV program watched solely by peasants is, in fact, a program that nobody watches.

The plague years

The inhabitants of Nucşoara and Scorniceşti have got quite a lot in common. In the first place, in both communes elections have been consistently won by the successor of the Communist Party. The prewar historic parties, the ones that everybody used to vote for in the two villages after 1946, won one round in the 1996 national elections, but gained no significant support in either village or in rural areas more generally. Most people in both villages claim that life was better under Ceauşescu. While in Scorniceşti the grave of Elena Bărbulescu was covered in wreaths, in Nucşoara an attempt by former political prisoners to place a cross in the cemetery in memory of anti-communist partisans fallen in the mountains met with determined opposition from most of the village. They had to retrace their steps, carrying the cross on their backs, and place it in a forest clearing just outside the village. The priest himself—the man with the satellite aerial—opposed the placing of the cross in the cemetery on the reasonable grounds that it would divide the community. Nowhere are such fears more justified than in Nucşoara, where the assets of those sent to prison were redistributed among the other inhabitants during the Stalinist years. The quasi-universal opinion, especially among those who were lucky enough not to have family members sent to the gulag, is that attempts by the victims to take justice into their own hands will divide the community. The law might have been on the victims' side, and, after ten years of court battles after 1989, newspapers in Bucharest might have been glorifying them or commiserating with them, but the village was against having the cross in the graveyard. This includes many families who, while not benefiting directly from the redistribution of the political prisoners' assets, nevertheless disapproved of the exceptional opposition of these people to communism, as the entire village suffered from the repression. The Securitate would indiscriminately load villagers from Nucşoara up in trucks, knowing that these people in fact had no information whatsoever on the partisans' hiding places in the surrounding mountains. Securitate officers would violently beat them before releasing them, thus ensuring preemptively that they would denounce the partisans should any information come their way. The partisans themselves would in turn threaten or even shoot at suspected police informants. The political detainees were amnestied in 1964 and, although the Securitate continued to watch them closely until 1989, there were never any problems in Nucşoara. And for good reason: After twelve executions, no-one would have risked it. Nearly forty years after the amnesty, justice was a secondary consideration at best and a liability at worst. All that people wanted was peace and quiet: Conflict was to be avoided, even at the cost of having past injustices revisited in the present.

How was life before communism? In Scorniceşti people were poor; in Nucşoara they had just enough land to make a decent living. They never had enough land to be comfortable and were obliged, especially in Scorniceşti, to do work for hire on the lands of the "boyars," people with fairly recent property titles, whose wealth had been considerably reduced by successive agrarian reforms. After the war, in its bid to attract and then subject the peasantry, the Romanian Communist Party employed strategies used similarly in other predominantly agrarian countries. In a first stage, via a new agrarian reform, the communists reallocated land to less prosperous peasants, whose gratitude they earned in this way. In Nucşoara, for instance, land expropriated from boyar Pavel Paul was allocated mainly but not exclusively to poor peasants returning from military service. In a second stage, the party gradually increased taxes and produce quotas, following up with a policy of requisitioning which was understandably unpopular.[9] This was a policy pursued in the Soviet Union in the 1920s and 1930s, in Eastern Europe after 1945, and in China in the early 1950s. It was practised in similar ways in all the villages and communes of Romania. There is, however, a basic difference between Nucşoara and Scorniceşti. In Nucşoara collectivization failed and people retained their land during communism, thus remaining in what shepherds used to call jokingly "free Romania" in the 1970s, under communism, when they had somewhat recovered their sense of humor. "Free Romania" meant the areas of unproductive mountain terrain which amounted to less than seven per cent of the country's surface area. Conversely, in Scorniceşti collectivization went virtually unopposed and was fully implemented. Those who resisted, and they were very few in number, were arrested and beaten, a story repeated in the remaining ninety-three per cent of rural areas. Each village thus typifies one of the two aforementioned categories, while retaining its specificity.

Indeed, owing to specific circumstances, from 1949 to 1959 Nucşoara was an exceptional village. So was Scorniceşti from 1969, the year of its new administrative and territorial reconfiguration, and up to 1989, when President Ceauşescu and his wife were executed. The consequences of this exceptionalism are still noticeable today. It was not so much the nature of the events as their particular intensity which set these two villages apart from other villages in Romania. Extreme communist practices in the two villages have left their imprint, and their outcomes are well worth studying.

Nucşoara was not the only village to serve as a base for partisan operations. The situation was quite common in mountain villages, but Nucşoara was the only village in which members of the resistance network were recruited from the highest-status group in the village, which ensured that partisans could survive for ten years in the area, despite being hunted down by the communists. This situation led to repression and punishment on an unusual scale. Likewise, in Scorniceşti, Ceauşescu's ambition to turn the village into a town led to successful systematization on an incomparable scale, the long-time effects of which can still be seen today. Before embarking on an analysis of the processes involved, it might be helpful to look at the local historical background.

Local history is mostly made of violence. At Slatina, Nicolae Ceauşescu, "a fearless young man"—to use the words of an admiring fellow-villager who wrote a panegyric of the communist leader in the 1970s—shot dead the director of a bank who refused to contribute to the communists' electoral campaign in 1946. He also shot into the air to intimidate the inhabitants of Tătuleşti, who were voting with the Peasant Party, amidst claims that the elections were being rigged. Those who did not vote with the communists were brutally beaten. The old school, where the polling station had been set up, was burnt down during the incidents.

In the other village, in the mountains, the area's most influential man, Gheorghe Şuţu from Domneşti, an entrepreneur and the uncle of the future partisan Elisabeta Rizea, was shot dead in his own shop and defenestrated. At Vîlsăneşti, a village neighbouring Nucşoara, a priest was assassinated. During the 1946 elections, the young Petre Ungureanu and his associates set fire to the ballot boxes, after which they had to go into hiding. And although the entire region had voted with the Peasant Party and the liberals, the communists were nevertheless declared victorious. Ungureanu claimed that, after this episode, he was captured by the Securitate and was in no position to refuse when they asked to be led to the partisans' hiding places in the mountains.[10]

As the army purges started in 1946, career officers were eliminated and replaced with under-officers promoted overnight or with former prisoners in Soviet camps, recruited from the "Tudor Vladimirescu" Division. This spelled the end of their career for Colonel Gh. Arsenescu and Lieutenant Toma Arnăuţoiu, who had both fought on the Eastern Front. The following year, 1947, brought the arrest and conviction of the leaders of the historic parties, and 1948 saw the nationalization of all farms, shops, mills, and rural businesses. Colonel Arsenescu, who had already organized two groups of partisans, was approached by a number of locals from Argeşul de Sus (Upper Argeş) areas who asked him to form and lead another such group in Nucşoara. They were all individuals pursued by the communist Securitate for "economic criminal activities and sabotage."[11] The members of the local economic elites felt trapped, as their life savings and assets were being threatened. They were people who had acquired wealth and had thereby "arrived," as the saying went about small fortunes accumulated no earlier than World War I. Iancu Arnăuţoiu, father of Toma—the young officer who became the leader of a partisan group—was a schoolteacher primarily, but also an entrepreneur who bought land, real estate, and cattle. He had acquired local boyar Iorgulescu's house by purchasing the latter's mortgage from a moneylender. Toma's father-in-law, Nicolae Niţă, was a mill-owner. Thus, the first people whom the Arsenescu-Arnăuţoiu group attempted to recruit were individuals from the rural middle class, the very social class which Ion Mihalache and Nicolae Iorga[12] had attempted to support and develop. Such people became the natural sponsors of the political refugees, as they sensed the net tightening around them and realized that they needed to build a self-defense base. The better-off farmers who had survived the 1945 reforms with fifty hectares of land represented a real problem for the communist regime, which lacked incentives to lure them into state cooperatives. Therefore, the regime requisitioned all their tools and equipment and gradually increased their taxes and their mandatory produce quotas. Even when broken by taxes—which went towards the repayment of Romania's war debt to the Soviet Union—the peasants were able to get their tools back from the state on loan and continued to farm and resist. Farm buildings in their turn were confiscated and later, once the "legal framework" for the requisitioning of fifty-hectare lands was finally in place, farmers were snatched from their beds at night and arrested. Arsenescu himself lost his farm, while one of the landowning boyars in Scorniceşti vanished one night never to be heard of again. He had been a rather decent man, and people remember their mothers crying as they told the story of how his family had built the old chapel, which was later demolished. Boyar Pavel Paul from Nucşoara was also arrested and tried alongside the partisans, with whom he had no links, simply because the communists wanted to confiscate whatever was left of his lands. In the mountains, there were additional reasons for tensions between locals and the regime. The mountain-dwellers had weapons, and hunting was an important occupation. Petre Ungureanu, the most influential man in Nucşoara, a former gamekeeper turned Securitate guide to help search for hiding partisans, had learned his hunting skills from Titu Jubleanu, one of the partisans. Jubleanu had gone into hiding because he had failed to give up his weapon before the prescribed deadline. Today, Ungureanu, a man who betrayed no emotion as he led the Securitate to the hiding place of two women partisans, still sheds tears over Jubleanu's Italian-made short-barrelled gun. His own arsenal, his pride and joy, which he allowed me to film with evident pleasure, was quite impressive too. During communism, those caught with undeclared arms were sentenced to between six and twelve months. Titu Jubleanu, an elderly man, together with his wife and son, were all killed. It was not simply their pride as hunters which prevented people from giving up their weapons, but also fear of reprisals and their instinctive independence as mountaineers. The state had seldom intervened in their affairs before. Peasants have been known to naturally resist any requests to give up valuables without compensation. A local peasant, brutally beaten up by the Securitate and asked to deliver his weapons, had replied: "But I don't have weapons, only a tank." Thinking that they were being mocked, the Securitate officers continued to beat him even more fiercely, until they understood: as the German army withdrew, it had left a small broken-down tankette in his orchard, which the man had kept hidden under hay. Under severe torture, he tried in vain to convince his captors that he had no intention of starting an insurrection with his tank, but only to use the engine to power his mill once the war and communism were over. He later shared his prison cell with others from the Nucşoara lot, even though he had never had any connections with the partisans.

Crucially, the local anti-communist resistance was encouraged by the illusion of forthcoming Western support. The partisans believed that they were going to withstand the abuses of the new communist regime for a few months, and then the West would intervene and things would return to normal. Radio Paris and Radio London—followed by radio stations such as Radio Free Europe, specially created for propaganda purposes during the Cold War—were the first to feed such illusions. Staff at Western embassies in Bucharest, too, encouraged such expectations, while inadequately trained agents were being parachuted into Romania by Western secret services to bolster the partisans' morale. King Michael of Romania once recounted how he personally tried without success to stop the parachuting of agents whom he knew were going to be executed on the spot by the Securitate. Western secret services tried to fulfill their quotas of secret agents by training former Iron Guard members or war prisoners. One of those thus parachuted, a peasant trained in Austria, was told by his family to forget about guerrilla resistance immediately and return to help with running the house. He was caught only because he foolishly went to the Militia station in the village to apply for ID papers.[13]

Colonel Arsenescu, a former staff officer, was not as naïve. He had stockpiled arms on his estate land near Câmpulung, as he had become convinced that Bucharest was no longer an appropriate hiding place and that Western embassies were no longer able to provide protection. He started distributing weapons, and also allocated pseudonyms, provided combat training and gave instruction on survival tactics. He was among those who realized quite early on that the arrival of Western cavalries was going to be delayed indefinitely and that, quite possibly, they might never come. As soon as his group started to be harassed by the Securitate, he found refuge with one of his former servants in Câmpulung. He lived in hiding there for years, before being spotted by a neighbor as he went to take some fresh air in the garden at night. He was denounced, arrested, and executed, sharing the same fate as the Arnăuţoiu brothers.

The villages thus provided the framework for the resistance networks, recruited on a daily basis from sympathetic groups of relatives, acquaintances, and in-laws. They were the eyes and ears of those in the mountains, as well as their chief suppliers of food. Toma Arnăuţoiu's mother sent the family's maid-servant into the mountains to look after her son, wash his laundry, and cook for him. The woman—Maria Plop by name—remained there to the end, had a child by Toma, and died in the gulag after they were caught.

The social gap between peasants with more land and better education and the rest continued to apply in the mountain hide-outs. The well-to-do were addressed as "boyar" and "Sir," although in Nucşoara there had been only two actual boyars, Iorgulescu and Pavel Paul. The differences among the other villagers had more to do with status than class. Status, as opposed to class, refers to a shared lifestyle rather than a common economic interest. Higher-status villagers, the village intellectuals, and the wealthy in their turn showed respect for the two genuine boyars, whose families had built their wealth comparatively earlier. Traditional authority and prestige-based hierarchies actively functioned in the Romanian village. On the lowest rung of the social ladder, servants were still recruited from the former Gypsy slaves, who had retained a sense of loyalty to their former masters. Two of the most active supporters of the partisan groups, Marina Chircă and her sister—who was already on the Securitate's radars—managed to hide for five years in the flea-infested loft of one Gypsy artisan, although the risk was high; the Gypsy man's wife, who had previously been in the service of the two women's families, had threatened to leave home if the husband did not offer them protection. All the intellectuals, that is all the priests and schoolteachers in the villages around Nucşoara, had been enlisted in partisan support networks. Village intellectuals had benefited on and off from prewar special policies aiming to create a rural professional middle-class capable in its turn of improving the peasants' education. Schoolteachers, for instance, had been the beneficiaries of special land allocations in the post-World-War-I agrarian reform. Iancu Arnăuţoiu himself and other rural entrepreneurs had prospered as a result of such measures. These special programs drew on a conception of social mobility shared by most Romanian populists, including Spiru Haret,[14] Nicolae Iorga, and Ion Mihalache. They believed that village intellectuals had to be examples of entrepreneurialism and civic virtue beyond their call of duty, as educators and civil servants. As such, they had to contribute to a process of moral and civic "elevation" of the peasant masses. These intellectuals, therefore, had been the natural support base of Peasant Party prewar governments, as representatives of an upwardly mobile peasantry and of the paternalism of a regime which would never allow itself to be duped by the communist rhetoric of "peasant exploitation." As they were comparatively well-to-do, they were subsequently lumped together with the "rich peasant" class (*chiaburi* in Romanian, or *kulak* in Russian, as they were frequently called by the party), and thus became doubly the enemies of the communist regime. Nine years into the Securitate's campaign against the absconding partisans, the agency streamlined its operation and started using undercover agents instead of mere recruits. Its newly found efficiency meant that the partisans were soon caught. All the intellectuals in the support network of the Arnăuţoiu group—the commune's teachers and priests, who had contributed food supplies and intelligence—were also arrested and executed alongside those whom they had tried to help. Among the first twelve members of the Arnăuţoiu group to be executed were three priests and two teachers. Other members and supporters of the group, however, had neither wealth nor education and were listed as shepherds or as poor and middle peasants in the court records of the communist so-called "people's" tribunals. Only the leaders of the group were registered as sons of well-to-do peasants. A middle peasant in Nucşoara today has around six acres (three hectares) of land, chiefly as a result of the perpetual fragmentation of the family plots through inheritance. The earliest recruits of the communist regime, people like Gheorghe Şerban, for instance, who came from a family with numerous children—a guarantee of poverty—had even less. Constantin Paşol, the other local communist, resentfully recalls that his family had only two acres of land and, as one of four children, he had to take up work for the boyars.

Nucşoara's partisans were never in a position to pose a threat to the regime, but in every direct confrontation they were speedier and better organized than the Securitate recruits sent to capture them. The only member of the partisan group with a pronounced inclination for violence, the medical student Ion Marinescu, shot dead two communist collaborators (Darie and Băncescu), new "cadres" actively working towards the communization of the village. These two had actually caused difficulties for the partisans' families. Other communists, such as Gheorghe, the mayor of Nucşoara, and Constantin Paşol, the first secretary of the local party branch, could have easily been attacked by partisans as they made their way to their isolated homes at night. This never happened, however, although later the Securitate attempted to frame those arrested by accusing them of involvement in an assassination attempt against Şerban. Like many others, the two communist notables sometimes ran across the partisans accidentally. Today still, Constantin Paşol calls the partisans "bandits," but admits that they were harmless. He himself had met Arnăuţoiu once and advised him to return to the village and join the communists. Such random encounters should not surprise. Before the Securitate started training undercover agents—whom it sometimes placed as lodgers in houses under surveillance—the refugees used to come down from the mountains into the villages around Nucşoara quite often. The group never had more than ten members hiding in the mountains, even at the time when Arsenescu and Arnăuţoiu were together, although the support network was much wider. Arsenescu was the only member of the group to have a well-defined sabotage plan against the regime and links with other groups, but he withdrew as a result of denunciations in late 1949 and lived in hiding until 1960.

The partisans had to split several times after Arnăuţoiu and Arsenescu each went their own way in 1949. The most enduring group was practically made up of two families: the two Arnăuţoiu brothers and Maria Plop on the one hand, and the Jubleanu family—father, mother, and son—on the other. In their last years in hiding, when it became obvious that they were no longer but a group of isolated, beleaguered people, their friends in the support network desperately sought ways of smuggling them abroad. The Jubleanu couple were caught as a result of an aerial reconnoitering mission. He gave in to her request to help the two "boys"—their own son and the young Marinescu—escape. As they were being chased, she did not stop when summoned and was shot dead on the spot. Her husband had to dig her grave there and then with his bare hands. Subsequently tortured, he gave up the locations of all of the group's reserves of food, which led to increased suffering among those left to face a harsh winter in the mountains.

As hope receded, the few partisans left fell prey to growing mistrust, and a fierce competition for the few remaining resources followed, which occasionally led to fratricidal murder. As the Securitate terrorized the families of those hiding in the mountains and those who might have had contacts with them, and as the net closed around them, members of the group had to restrict their activities to hunting for food, clothing, and other basic supplies. Survival, rather than fighting the regime, became the rule of the game. Maria Plop had a miscarriage, and then gave birth to a second baby with only the Arnăuţoiu brothers assisting the birth. When she was captured, she was suffering from tuberculosis. When the Arnăuţoiu family were finally arrested, all those previously convicted—i.e. practically all those arrested beginning with 1949—were given a retrial under more serious accusations.

Partisans from other regions such as Bukovina, who escaped execution, recall the terrible deprivations of those years in hiding. They could bathe only in the summer, and they avoided using the washing facilities in their hosts' homes for fear of being surprised and captured. For the same reason, they would always sleep with their shoes on. They had no medicines, and a mere pneumonia could kill them. Their capture—although followed by further interrogations and arrests—was met with a collective sigh of relief. Because they had scrupulously written down every item which they had borrowed or requisitioned from the villagers, from cheese to maize and cigarettes, these lists led to over sixty convictions and over forty confiscations of assets, which benefited other villagers. People in the neighboring villages were relieved. The situation had become desperate. Some went to the Militia—which had its headquarters in the former Arnăuţoiu parental home, purchased from boyar Iorgulescu—to report accidental meetings with partisan group members as they went to check on their sheep or to gather wood. Such people were beaten up brutally either in order to be "encouraged" to report on further such encounters, or to make sure that they remembered every detail of them. But above all, they were beaten to serve as an example to the entire village community and to create a state of total terror.

Those who survived interrogations and prison were liberated in the general amnesty for political prisoners in 1964 and were sent by the communists to work as foresters or as gamekeepers. Their workmates were supposed to keep them under observation, and many of them did, up until the fall of the communist regime in 1989. Former detainees, too, had been instructed to keep each other under surveillance, but they had forewarned each other and were cautious. Until they were able to have new houses built, many lived as lodgers in their former ones, now belonging to others. Vică Berevoianu from Nucşoara recalls even today how one night, helped by relatives, his wife moved all the building materials destined for their new house over onto her own land to avoid confiscation. He explains that the lands of wives who did not have to go to prison (although many followed their husbands there) were not confiscated. Berevoianu's wife earned his forgiveness for having divorced him while he was in prison as she was able to prove she was acting in their best joint interests. Although they lived together again after he was amnestied, they remarried only after 1989, when they were both over eighty.

The fate of the children of the two protagonist couples was even more tragic. The child of Maria Plop and Toma Arnăuţoiu, aged two when they were captured, was put up for adoption through the kind services of a doctor from Domneşti, and lived under a different surname and even a different name. Colonel Arsenescu's wife was arrested, and his son, too, changed his name. When the Arnăuţoiu brothers were captured, the People's Tribunal held a retrial for all those captured after 1949, when the Arsenescu group had been caught. In the light of new evidence, they were all executed, including an undercover Securitate agent, Benone Milea, who could not be saved even by his police liaison officer's testimony. The court heard that Milea had used his bayonet to finish off a fellow Securitate agent fatally wounded in an ambush, probably in order to earn the trust of a suspicious Arsenescu. The tribunal did not believe the Securitate's claims that Milea was an undercover agent and, as he had not sent many written reports, it was believed, possibly accurately, that he had joined the partisans for good. He was executed after many years in custody pending an appeal which he lost.

The invention of the new countryside

The special story of Scorniceşti and its people started in the 1970s. By then, Nucşoara's political prisoners were either dead or had returned from the gulag to a village which in the meantime had become a hostile environment. Up to that point, the history of Scorniceşti—the poorer village—had been no different from those of other collectivized villages of the Romanian plains, although its famous son, Ceauşescu, Minister for Agriculture and the most active champion of collectivization, had managed to drag it prematurely along the path of communization. A paradoxical project with no equivalent elsewhere in Romania aimed at modernizing Scorniceşti by turning it into a town through radical "systematization" on the one hand, while still upholding it as an example of rural tradition and culture on the other. While all the houses in the village, starting with those owned by people who had resisted collectivization, were being demolished and replaced with high-rise apartment blocks, Ceauşescu's parental home and the village cemetery were given an "authentic" makeover with "artistic" touch-ups. A new museum and cultural center were built to house the work of popular artists from all over the country, including Bucharest.

Tătărăi appears to have been an exceptionally poor village: People's testimonies, older monographs, as well as literature produced under communism all seem to agree on this point. The village had no artists, not even the usual potters or the artisan carpenters who insisted on embellishing the utilitarian objects they produced. The area boasted a colorful traditional dance called "Căluş," but nobody in the village itself actually practised it. "The traditional Căluş was an import. It was [the tradition] in Pădureț, Poloneț and Mogoşeşti, but not here in Scorniceşti. The dancing went on from one end of the village to the other for a whole week at Whitsuntide. Groups from various villages took part in what became a Căluş competition. But Scorniceşti was not part of this tradition." (Scorniceşti graduates focus group)

Scorniceşti was constructed from scratch as a model village of hard-working, artistic folk people. The massive gates of the cemetery, carved by a regional artisan in the traditional style of sub-Carpathian Oltenia, are a testimony of those times. Nowadays, there are no carpenters left either in the small town or in the adjoining villages. Even a modest, ready-made wooden cross of the kind placed on a fresh grave before a permanent stone memorial can be commissioned and has to be purchased in the nearby city of Slatina. In the past, the museum had artists-in-residence who gave demonstrations of pottery-making and embroidery. But these artisans were not locals, and today both they and their artefacts have disappeared without trace. The museum, which housed more Ceauşescu family memorabilia than ethnic artefacts, was vandalized during the 1989 Revolution. All that is left are a few skeletal display boards covered in red velvet, now bereft of the old communist symbols and slogans. The entire collection of invented artisan craft has otherwise vanished.

Today the building, which also survived a fire, hosts the public library and a couple of classrooms. The library now holds only some twenty volumes of Ceauşescus' works, which used to comprise hundreds of volumes. The remaining copies are mostly brochures spared by the devastation. Otherwise, it is a typical rural library, with titles ranging from the constructed histories of Romania written under the patronage of nationalist communism—but still used by history teachers—to translations of world classics into Romanian commissioned under the communist regime. Cash-starved public libraries today can hardly afford to buy new books. As translators during communism were frequently intellectuals of the old regime elites marginalized by the communists, they had to publish under a pseudonym, and the translations—ranging from Dostoevsky to Gide—were of much higher quality than translations produced later. However, the library's records show that the students of the agro-industrial college are not exactly voracious readers: A best, they read the titles on their college-recommended reading lists. After all, they do not need reading: the more advanced Scorniceşti acquired cable television ahead of Nucşoara, and shortly afterwards an Internet dealer too. The folk dance group "Căluşul" was the only such group to survive the 1989 Revolution, although the dancers requisitioned from Bucharest have since left the ensemble. Nelă, the cultural entrepreneur, still gets annual funding for his company from the local budget. Under Ceauşescu, they used to tour the world as representatives of Romania's national arts. Nelă watches over his dancers' diet and makes sure they are fit enough to jig. When the stars left, he started recruiting from among the local talent. The slightest allusion to the particular circumstances which ensured the group's success makes him jump:

> We represented our local community and Romania on three continents: Asia, Europe, and America. We won awards everywhere. We were the best by far, the only serious competition came from Bul-

> garia, Yugoslavia, and the Russians. [...] We were invited to high-level festivals by presidents and state leaders, Elizabeth II, Queen of England, King Hussein, not just me, but the whole group, we were well-received, we gave gala performances, and not because of Ceauşescu, but because of us, because we had this gift from God.

The history of Scorniceşti was the subject of a monograph by one of its natives, Ion Spălăţelu, who became a teacher of history in the capital, Bucharest. His focus was on the role of the Ceauşescu family, and he emphasized the national character of folk art—an emphasis which coincided with a general ideological shift in Romania in the 1970s and 1980s. At that point, Ceauşescu's speeches started to refer to prewar historians, and national history was being rewritten not only from the angle of class struggle, but also from a nationalist perspective. At the Romanian Communist Party's Museum in Bucharest the statues of Marx and Engels guarding the entrance on both sides were relocated to make way for the Roman Emperor Trajan and the Dacian King Decebal, the official founding fathers of the Romanian people, in whose direct lineage Nicolae Ceauşescu placed himself. In the 1980s, investment in folk culture served the regime's plans for mass mobilization. Its outcomes were the mass cultural event "Cîntarea României" [Song to Romania] and the creation of the Front of Socialist Democracy, an organization designed to include all the politically non-affiliated, pensioners, trade union members, and people who were not members of the Communist Party. While the bulldozers were ploughing their way through the historic center of Scorniceşti and other villages, the activists in charge of the "Song to Romania" claimed that they were preserving the heritage of traditional folk art. They did indeed manage to mobilize people in their hundreds of thousands. State television produced county-by-county series of the festival's history, and the cultural activists mobilized everyone, from artisan potters and seamstresses to students and teachers.

It was not solely the relocation of its inhabitants into apartment blocks that turned Scorniceşti into a town. The commune was also endowed with a factory of fine mechanics, a textile factory, a brewery, a slaughterhouse, and gigantic solaria. Intellectuals and professionals who wanted to relocate there were fast-tracked to obtain accommodation in the new blocks which, in the beginning, had both hot and cold running water. A huge hospital was built and staffed with medical personnel from Slatina. A mega-cooperative farm was also built, the largest in the country, complete with an extensive animal farm. Ion Neacşu, the village's communist mayor at the time, was immensely proud: "We have built stables and filled them with cattle," he declared. Perfection was attained when, apparently without consulting with Ceauşescu, his brother-in-law, Lică Bărbulescu, a football fan, had a huge stadium built on the model of the Roman arena in Verona. It had a capacity of 20,000 seats, way above the commune's total population, newborns included. The rim of this ring-shaped structure is formed of covered spaces, initially conceived as manufacturing and commercial units to justify the absurd building cost. The local football team, formerly a fixture of the county league, was transformed overnight into a premier league football club, F.C. Olt, with players recruited nationally. Referees were notoriously reluctant to penalize a team which symbolized the Ceauşescu family, and playing against F.C. Olt was a nightmare for other teams. Naturally, the club fell apart at the collapse of the communist regime and now the stadium stands empty. However, the enterprising mayor rented the circular space to textile manufacturers, which organized sweatshops based on importing fabrics and designs and on paying minimal wages. Staff are paid rates below the minimum national income. The women and youngsters who work here make around fifty euros per month, but at least they are employed. All around them, state manufacturing industries are in free-fall. The small town is, therefore, happy with its sweatshops, which employ practically the entire local female population. Lorries displaying the labels of big German and British clothing industry firms come to collect the finished goods. They lumber along the village "high street"—built when the place was "systematized"—giving priority to rows of fretful geese, women carrying sackfuls of produce on their heads, and men walking along the pot-holed sidewalks with scythes on their back. Countless stray dogs, left homeless after successive demolitions, trot along the same busy main street, now half colonized by the textile firms. Teenagers—those whom Nelă failed to recruit into his dance company—keep fit by dancing to rap music on the terrace of a tavern. The name Ceauşescu is meaningless to them. They are being watched by girls with baseball caps worn back-to-front in *Beverly Hills 2001* style and by boys who look like MTV clones. Courtesy of the firms which employ their mothers, the youngsters of Scorniceşti can afford to dress like their Western counterparts. History, whether distant or recent, is no concern of theirs, while folklore is just a reminder of the unfinished emancipation of their parents from their peasant status—something they are not keen to recall, eager as they are themselves to outgrow it.

Notes

[1] The Gypsies (also known as Bohemians or Tsiganes, and, more recently, Rroma) were slaves in the boyar households of Wallachia and Moldavia. Gypsy slavery was phased out gradually from 1837 to 1856. Estimates of their total number today in Romania vary from around 600,000 to 2,000,000.

[2] Tudor Vladimirescu (1780–1821) was the leader of a popular short-lived uprising directed against the Sultan-appointed Phanariot rulers of Wallachia, one of the constituent principalities of Romania.

[3] During World War I, Romania remained neutral until August 1916. After a promising start alongside the Entente forces, the ultimately isolated Romanian army was defeated and half of Romania's territory was occupied by the Germans. The Allied victory reversed the situation and placed Romania on the victors' side, which allowed for the political unification of the historic provinces: Transylvania, Bukovina, and the Banat, formerly part of the Austro-Hungarian Empire, then joined the Old Kingdom of Romania, formed out of the union of Wallachia and Moldavia in 1859.

[4] After the installation of the communist regime in 1947, pockets of armed resistance were created all along the arch of the Carpathian mountains. They were mainly groups formed at the initiative of former army officers and former members of the legionnaire movement. These groups, isolated from each other, and comprising around ten members each, operated on the basis of regional networks built around links between their mountain hide-outs and the surrounding villages. There were around fifteen such armed, but uncoordinated, groups which were finally defeated by the Securitate and the army. Their repression ended in the 1950s when members of the groups were either killed or sentenced by military tribunals and executed. The anti-communist resistance movement in Nucşoara is the most famous, owing to the two volumes of documentary evidence collected by Ioana Arnăuţoiu, daughter of one of the movement's leaders, and by the Romanian Centre for the Study of the Securitate Archives, as well as to large amounts of oral testimonies.

[5] Blaga, "Permanenţa preistorică."

[6] The dreaded communist secret police.

[7] The main European Union aid programme for associated countries.

[8] CURS poll, 1999, as reported in Mungiu-Pippidi et al., "In the Shadows."

[9] See Seton-Watson, *The Pattern of Communist Revolution*, 345.

[10] Video-taped interview by Alina Mungiu-Pippidi, June 2001.

[11] See Voicu-Arnăuţoiu, *Luptătorii din munţi*, 48.

[12] Nicolae Iorga (1871–1940), Romanian writer and politician. He was a prolific historian, poet, journalist, and dramatist who promoted national-

ist and rural themes in Romanian literature. As Prime Minister between 1931 and 1932, he was assassinated by members of the Iron Guard.

13 This case has been documented by Marius Oprea of the Romanian Institute for Recent History (Institutul Român de Istorie Recentă).

14 Spiru Haret (1851–1912), Romanian mathematician and politician. As Minister for Education between 1897 and 1910, he created a national education system and encouraged the expansion of education in rural areas.

CHAPTER 3

The Construction and Deconstruction of Rural Property

Peasant land tenure in historical perspective

Rural societies have no memory. Seasons of their life are very much alike, and peasants do not indulge in record-keeping: only the propertied classes do. Consequently, no-one in Nucşoara or in Scorniceşti can look back beyond their grandparents' generation, and when they do the resulting picture is rather blurred. The oral history of Romanian peasantry thus starts with the aftermath of the agrarian reforms of World War I. These were the most comprehensive reforms of their kind in Europe, due to the need to breach the widest gap on the continent between a small group of big landowners and a huge mass of landless peasants. Some scattered memories do remain of the old 1864 land law instituted by Prince Alexandru Ioan Cuza,[1] who nationalized the Holy Land monasteries in Romania (which in Wallachia held about a quarter of the entire acreage of the country and in Moldavia almost one-third, and which in the nineteenth century still sent their dues to Sinai, Mt Athos, and Palestine) and distributed their land among the peasants. Those who did not benefit from that reform rioted in 1907.[2] In Nucşoara, where the terrain is mountainous, there are a few properties dating back to Cuza's time, but most peasants received the lands, they still own today, after World War I. Only in the village of Călineşti-Argeş, at the manor of boyar Gheorghe Paul, son of Nucşoara's chief landowner, do family portraits go back as far as the eighteenth century. In Scorniceşti I could find no-one whose land had been acquired prior to 1917.

Community joint land tenure, the most archaic form of landownership in Romania, is not within the living memory of Nucşoara's inhabitants, and was unknown in Scorniceşti which developed late as a village. Historically, joint land tenure gradually disappeared under pressure from Ottoman domination and Western grain markets, which demanded a more rigorous and systematic type of land management. In Central and Eastern Europe, serfdom survived the demise of its Western counterpart by a few centuries: The enlightened Prince Constantin Mavrocordat formally abolished it in the mid-eighteenth century. However, the freed peasants, whose only capital was labor, remained dependent on the large landowners and the monastery estates until Cuza's reforms a century later. In contrast to parallel developments in Western Europe, this was not a period of industrialization and urbanization in Romania. In addition, although demographic density increased—without attaining West European levels—farming remained the chief source of surplus, and the vast rural areas were the ones supposed to accommodate everybody. Landholding peasants, amounting to around twenty per cent of the population in each of the Romanian Principalities, had been in possession of their own plots for several hundred years, yet they never became a class of small landowners and continued to live at subsistence levels. One explanation for this was the low level of technological advance in farming. Agriculture was still practised with primitive means and, as the population had been rather sparse historically, there had been no incentives for improving productivity until the Peace of Adrianople in 1829,[3] which temporarily turned the Romanian Principalities into one of Europe's chief suppliers of grain. This applied to the entire region. In Romania, however, an additional factor accounting for the persistence of subsistence farming was the state's inability to offer an alternative for absorbing the offspring of landowning peasants into the army, the Church, and the cities. The fundamental legal institution which led to the consolidation of a class of petty and middle landowners in England, for instance, was the right of male primogeniture. This did not exist in Romania, where all descendants continued to exploit or live off the land in common. This model survived to modern times. The positive opportunities offered by both Cuza's reform and the reforms at the end of World War I were wasted because of the peasants' practice of dividing lands among heirs. Within the span of a single generation, this practice led to a shortage of land within the family and to the need to perform work-for-hire on large landed estates. Never, in the entire history of land management in Romania, was an increase in productivity popular as a viable solution for the growing need for cultivated land, though government policies in the first half of the twentieth century did try to promote policies encouraging it. The only lands where productivity increased were the boyar estates, but those were mostly liquidated by the post-World-War I reforms, which led to the collapse of Romania's grain export. In 1913, eighty-three per cent of Romanians worked in agriculture, and per capita income was among the lowest in Europe, on a par with Serbia's and Bulgaria's. Productivity in agriculture was constantly below the European average.[4]

The interwar period was particularly difficult for the peasantry, although the agrarian reforms first and the waiving of farmers' loan debts later showed that the political will existed to ameliorate the condition of farmers. In 1930, 71 per cent of the Romanian population lived—badly—off farming, compared with 8 per cent in England, 31 per cent in the Netherlands, 33 per cent in Czechoslovakia, and 51 per cent in Hungary. The contribution of agriculture to national income had fallen from 60 per cent in 1913 to only 53 per cent in 1938, yet it remained high compared with Hungary (36 per cent) and Czechoslovakia (23 per cent), and was comparable only with Bulgaria and Yugoslavia (53 per cent). Even today, the same development gap divides the Central European from the Balkan states.

Prior to the creation of Greater Romania with the addition of three new provinces in 1919, in the Old Kingdom (Moldavia and Wallachia alone) the large landowners (i.e. 5,385 individuals, amounting to 0.64 per cent of the total number of landholders) held almost half of the total land. The remaining 900,000 individuals owned only 52 per cent. The average area per individual landowner was 70–75 hectares, and per peasant family only 5 hectares. Extensive land redistribution was offered in order to ensure the loyalty of conscripted peasants during World War I, when the Romanian state was almost dismantled after being defeated by Germany. Almost four million hectares were thus allotted to one and a half million peasants, which led to the de facto demise of landowners as a political class and the decline of agricultural production. By 1940, one-fifth of the households owned only one hectare of land each, a plot too small to ensure a family's subsistence given Romania's productivity levels. The *Enciclopedia României* of 1942 itself explained the low productivity as arising from the peasants' need to employ the labour of all family members, who lacked other opportunities.

This typical arrangement could be found at Scorniceşti, in the Bărăgan Plains, but in the mountainous location of Nucşoara, the situation of landowning peasants was better. As mountain properties comprised herds and pastures, more people could live off one hectare. It was deemed that a smaller family could survive on two hectares in the lower plains regions, and on just one in the mountains. After World War I, what was left of the great landed estates around Scorniceşti was expropriated. Olt county had its fair share of old boyar families: The Brâncoveanu, the Callimachi, the Blaremberg, and the Soutzo were among the great Romanian landowners in the region. Overall, 63,450 hectares of land were expropriated in the county, of which 52,000 were distributed to the peasants. The reformers also allocated 8,600 hectares to common pastures and animal husbandry. In Scorniceşti itself, the largest expropriated estate was General Alexandru Lînaru's 178 hectares of land.[5]

Rural overpopulation remained a major problem for interwar governments. The peasants continued to work their own lands mainly as unpaid agrarian workers, and subsistence farming remained the rule for most families. Several interwar governments addressed the intractable issue of peasant dependency while also attempting to relaunch Romania's economy, with limited or controversial results. Even though the majority of peasant households have never become economic success stories and survived chiefly from state handouts and debt cancellations, by 1945 most peasants managed to live off their lands. They did not become prosperous, but neither were they as impoverished as they had been before World War I. The literacy rate improved dramatically between the two world wars.

The World War II campaign against the Soviet Union, when Romania joined Germany in the march to Stalingrad, led to huge losses among the male population: the majority of the 350,000 dead on the battlefields, were peasants. Others were taken prisoners and recruited into the "Tudor Vladimirescu" army division to fight alongside Soviet troops on the Western front. War on the Eastern front contributed considerably to the installation of communism in Romania. By the time the army had retreated back from the river Don, the number of the dead, wounded, and disappeared totaled over 600,000, more than wartime dictator General Antonescu's initial mobilization of half a million men.[6] Alarmist rumors aside, the survivors were fully aware of what a country could expect from a Soviet occupation, and they became the core of the irregular army of partisan-peasants who fought against collectivization. Toma Arnăuţoiu's elder brother, Ionuţ, had fallen on the Eastern front. An entire generation of officers, Romania's best military staff—recruited mainly from wealthy peasants, for whom joining the officer corps was the handiest road to upward mobility—were discharged from the army because they had fought against the Soviet Union. Arsenescu and Arnăuţoiu were among them. On the other hand, upon liberation poorer peasants taken prisoners by the Soviets became raw material for the agents of communization in the village. Such was the case, for instance, with Gheorghe Şerban, Nucşoara's first communist mayor, who had spent four years in a Soviet camp.

Interwar policies failed to solve the problem of rural overpopulation. The war proved once again that the only real surplus in an underdeveloped and non-democratic country was human life, but even war failed to solve the problem in its entirety. However, overpopulation quickly became a secondary issue. In spite of everything, peasant life improved considerably between 1917 and 1945. As a result of the most recent agrarian reforms, the peasantry now had more advantages to defend than to gain. The transient industrial boom of the prewar years, when protectionism and state investment had pushed Romanian industrial output to previously unattained levels, also affected lifestyles in the countryside. The wireless radio appeared—at least one in every village—and home-made or locally produced goods were being replaced by industrial goods, now cheaper and more easily accessible. As a result, by 1945 the revolutionary potential of Romania's peasantry was at an all-time low.

Legal terror and state violence

Aware of all this, the Romanian communists—brought to power by the Red Army—initiated policies aimed at placing the peasant class in a state of dependency. Their strategy involved a land reform whereby the large landed estates which had survived the interwar reforms would be reduced to fifty hectares each. Agitation against major landlords started prior to 1944, through manifestos which encouraged the peasants to occupy the large estates. These included assurances that the lands were not to be collectivized either then or later, but placed directly "under the individual ownership of endowed peasants."[7] The peasants were in a state of agitation and were unanimously opposed to the rumored imposition of the Soviet model. An overwhelming majority of the peasantry were in favor of the solutions advocated by the "bourgeois" Peasant Party as they feared that, if deprived of their lands, they would end up queuing at soup kitchens. Although their so-called political autonomy in the interwar period is questionable given that rural votes were always being manipulated by agents of influence, their economic autonomy had increased considerably in the last decades. Even in the realm of politics, patronage in the countryside did not amount to a total annihilation of freedom, which was the case in the Soviet Union under the communists, and the peasants were fully aware of this. Peasants voted according to their patrons' suggestions not because they did not have the freedom to vote otherwise, but because they exchanged votes for favors.

The 1945 reforms cut the maximum permitted area of rural properties to fifty hectares. It did not affect the wealthier peasants, only the larger landowners. In Nucşoara, the worst hit were the Paul family, and in Scorniceşti the Lînaru and Ledelea families. The main beneficiaries were landless peasants or those with lands under five hectares. In Nucşoara, this category included several World War II veterans, who were endowed with land from the estates of Pavel Paul in Slatina, one of the commune's component villages and later the site of a state farm. Given the fragmentation of mountainous plots, the former boyar's estate was the only one which lent itself to such an experiment.

Any illusions the peasants might have held were soon dispelled. The following year, 1946, saw a terrible drought, and new legislation (passed on 16 February) imposed new taxes on peasants and obliged them to sell their products to the state at fixed prices. The local cooperatives and even the army were entrusted with the collection. Industrial goods would be supplied to the countryside only if villages managed to meet the targets set for such mandatory sales. Subsequent legislation further restricted peasant control over their own produce. Thus, in Law 251 of 1947—which governed the circulation of agricultural produce—the Ministry of Trade established the type, quantity, price, time, and place of delivery to the state (informally known as "quotas"). Subsequently, Decree 112 of 1948 created a State Commission which oversaw the collection of grain harvests. In 1949, Decree 183 regulated penalties for economic crimes. The destruction, stockpiling, and adulteration of goods, for instance, led to sentences of up to fifteen years' hard labor. Those who failed to implement the state plan were liable to prison sentences of between one and twelve years. The system thus established a legal framework whereby the peasants no longer owned their own produce and, in addition, were liable to severe penalties if they attempted to manage their lands independently. At the same time, there was large-scale requisitioning of agricultural equipment and livestock. For Marinică Popescu, a middling peasant from Scorniceşti, it all started to go downhill the moment he lost all his farming tools, seized by the state farm (GAC) created in the vicinity. Subsequently, he was told that he was not allowed to farm his land unless he joined the state farm. Because at that time he was working as a clerk in a savings bank, he was told that he was going to lose his job unless he brought his land into communal ownership. This was a loose type of collectivization, designed to get the peasants used to the new system without, however, scaring them completely with the specter of collectivized production. So, he joined. He was later summoned to bring his cattle. He kept a horse, which he later sold at the market to get cash to repair his house. Nevertheless, the state farm claimed the horse and withheld 1,500 lei from his salary, a huge sum for the time, which he had to pay in several installments.

Once the penalties were in place and the agents of communization in the villages had thus acquired a handy leverage system, the state started changing the rules of the game as it went along, increasing the obligatory quotas for instance. In addition, these quotas were progressive and aimed in the first instance at ruining the large landowners. This system was extended from grain production to other produce, and eventually managed to bankrupt individual households. People who owned over five hectares and two cows were required to supply over 220 liters of milk, and those who had over eight hectares had to supply milk in direct proportion to the number of hectares they held (up to seventy-five liters per hectare) even if they did not have enough cattle. Gheorghe Popescu from Secături-Nucşoara was thus forced to keep four cows, although he did not have sufficient pasture-land, and had to take loans to buy milk in order to deliver his quotas. Taxes increased, and in 1951 people were required to pay them in advance. Kulak households and farms which had not joined the state had the doubtful privilege of paying taxes of 100 per cent and above on the sale of their produce. Similar to Russia, the state introduced a ban on the production of plum brandy, one of the most popular enterprises among peasants. In the mountains and the plains alike, the peasants used to make plum brandy for their own use, not for sale. Popescu, and many others, carried his distillation tank and hid it in some ravine, up the "devil's stream," as it was known locally. The brandy barrels and casks were then buried, away from the prying eyes and hands of the "quota agents."

Apart from the erosion of their rights to dispose freely of produce from their own lands, the peasants also suffered from gradual restrictions on the sale of land itself, which by 1947 practically destroyed the land market. The communist regime then proceeded to the de facto abolition of private property. On 2 March 1949 the state expropriated all agricultural enterprises and model farms, which were lumped together on the same list as the large properties. The pools of machinery and the buildings which had not been confiscated already were seized along with the land on which they stood. If the owners protested, they were penalized in accordance with Decree 183 and were given sentences of up to fifteen years' hard labor. Landlords who had not taken the precaution of hiding in their houses in the city or with family and friends were often arrested at night and deported. Others, such as the astronomer Nicolae Coculescu from Scorniceşti, whom the peasants liked because he warned them of bad weather by sounding his horn on the balcony of his manor, had prudently sought to anticipate the confiscation of his estate by an act of donation to the state. The Coculescu family possessed an eighteenth-century manor in Constantineşti, which they donated in 1948. It became the head office of the kolkhoz. Another wealthy landowner from Scorniceşti was not so lucky and was arrested, "taken away"—a peasant euphemism for being sent to the gulag. "Being away" quickly became in peasant parlance "away at the Canal"[8] or in some similar place. Ion Gavrilă, the chief of the partisans in Făgăraş, was to learn the meaning of the new phrase from a peasant woman he met on the road. As he had been hiding in the mountains, he had not been involved in the creation of this new vocabulary. The Paul family from Nucşoara initially survived, although their house in Brădet was confiscated and later converted into the village dispensary. They were a popular family, who had made their fortune after Cuza's reforms and had given a lot of support to middling independent peasants in the region. Today still, the villagers remember Pavel Paul as a model farmer, who would inspect the estate on horseback and help peasants with loans of hay for indeterminate terms or with whatever else they needed to survive. The communists had to fabricate some slander regarding his alleged complicity with the partisans in order to have him arrested and confiscate the remains of his estate, while all around the region small properties survived.

Collectivization progressed gradually, very slowly to begin with, as shown in the transcripts of the agonizingly long debates of the party hierarchy, and with occasional respites. However, under the guidance of Soviet advisers and with the cooperation of the communist militias, the process progressed inexorably, helped along by incentives such as tax exemptions for those who joined the kolkhoz, and by expert combinations of penalties, quota systems, taxes, and bullying. In theory, joining remained "voluntary," as shown in the party records and the statutes of the state farms, but the figures suggest the scale of the mounting repression. Thus, in 1961 a report to the Central Committee of the Romanian Workers' Party (PMR)—written by its First Secretary, Gheorghe Gheorghiu-Dej—shows that 30,000 out of 80,000 arrested peasants were subjected to public trials for failing to fulfill state obligations.[9] That figure was out of all proportion to Romania's population at the time. For comparison, in the Soviet Union in 1929, the year when repression and preventive arrests peaked, there were 100,000 arrests among a peasant population ten times the size of Romania's.[10] The repressive methods used included "arrests, beatings, torture, threats at gunpoint, ideological pressurizing during night-time summons in front of the Provisional Committee, forbidding children access to school, and scaremongering among middling peasants, who feared being placed on kulak lists."[11] In the early days of collectivization, the start-up land rarely came from members' properties, because the latter were recruited mostly from among impoverished, marginal peasants. The land, therefore, had to be expropriated from someone else. Expropriations were usually ordered under the pretext of economic sabotage allegedly committed by landowners. If peasants rebelled, which happened in dozens of villages, they were executed on the spot, which was the case during uprisings in Bihor county in 1949. The milder alternative was deportation to the wastelands of the Bărăgan, which was the fate of many in the villages of Transylvania and the Banat. Repression was accompanied by drastic economic sanctions, such as imposing low prices for grain and livestock, so driving producers into bankruptcy. Decree 115 stipulated the "abolition" of the last surviving "forms of man's exploitation by man." "As a result of the achievements so far in the construction of socialism in the countryside and of growth in the socialist agricultural sector, the abolition of the last remaining forms of individual farming, strip farming and tenant farming, and of all forms of man's exploitation by man in agriculture has become an *objective necessity*" (emphasis added), the decree announced. Article 10 stipulated that land farmed communally would be transferred to state farms "without compensation," and Article 11 stipulated that those who opposed this measure would lose all their land, even the land farmed individually.[12] Households were thus pushed to the limit of subsistence. The year 1959 marked the final assault on rural property, when voluntary collective tenures were transformed into kolkhozes.

Table 1. The progress of collectivization

Year	Kolkhoz	Voluntary associations	Hectares
1949	56	–	14,692
1950	1,027	–	277,719
1951	1,089	–	301,690
1952	1,795	1,834	924,008
1953	1,997	2,026	1,061,807
1954	2,070	2,833	1,199,313
1955	2,152	4,471	1,324,632
1956	2,564	8,130	1,855,002
1958			1,760,000
1959			8,400,000 (including the state sector)
1962			10,231,000

Source: Iancu et al., *Colectivizarea agriculturii în România*

Fourteen years into the collectivization campaign, in 1962, the Communist Party's First Secretary Gheorghiu-Dej was able to announce triumphantly that 3.5 million families of private owners had become members of collective farms. In the process, they had lost not only their land, but also their tools and livestock. The extreme violence unleashed in the campaign by promoters of the program was ultimately blamed on the "right-wing deviationists" in the party, more specifically on the Teohari Georgescu-Ana Pauker group, locked for a long time in a power struggle with Dej.[13] They were also blamed for the fact that the annual quota of requisitions was not met and the rhythm of collectivization was too slow. Nobody seemed to notice the glaring contradiction between these two statements. The minutes of top-level meetings of the Romanian Workers' Party (PMR) show, however, that it was not Ana Pauker but a much larger group which continuously promoted policies of violent repression against the peasants. This group notably included Nicolae Ceauşescu who was involved from the start in the activities of the first commission for collectivization. Dej commented with hypocritical disapproval on the methods used by the group: "And what were the methods employed? Torture, guns, etc. And whence come such methods? Were they not suggested by yourselves? Were they mentioned in the documentation? I personally have seen no document which says: 'Catch them, put them on trial, take their lands, "dekulakize" them, etc.' Such pressures may differ from one region to another. The documentation shows that such pressures were exercised in all counties, which means that this method must have originated somewhere. And if it did not start from us, from the center, it must have started from another centre."[14]

The mysterious "other center"—as in the famous SF novel by Isaac Asimov, *The Second Foundation*—was in fact none other than the first. The same report on collectivization so severely criticized by Dej admitted to the mass scale of show trials. It was only once they acquired "mass proportions that the Commission asked for such trials to be no longer admitted and that the Ministry of Justice decided to have the evidence examined. For instance, there was one case of someone being sentenced because he did not contribute his share of quotas deliveries, but these had not yet been enacted at the time."[15] Soviet advisers, such as "Comrade Veretennikov," were often placed in a position where they had to moderate the zeal of the Romanian communists, who tended to copy the Soviet model indiscriminately, including its obvious errors, such as compelling peasants to sell their stocks of seeds in order to pay taxes, and thereby reducing them to starvation. Accused by Dej of harming the party's image by employing unduly harsh methods such as "torture and guns" against the peasants, the First Secretary D. Nagy blamed the Securitate, the political police, which, he said, acted separately and followed only its own orders.[16] He admitted, however, that both the Militia and the Securitate were in fact under his orders, and that he had failed to double-check their "work." That the repression was in fact carefully coordinated is suggested by the large-scale use of the Militia, whose troops were theoretically supposed to protect civilians from crime and abuses. To an even greater extent than the Securitate, the Militia troops harassed the partisans, arresting those who opposed collectivization, and created a state of terror by using torture to force villagers to denounce absconders. The torture and beatings which First Secretary Dej openly admitted took place routinely in the offices of the local council, at the village hall. The mayors, including Gh. Şerban in Nucşoara and Florea Cârstea in Scorniceşti, kindly loaned the premises before making a discreet exit, and keeping out of it. Yet, it was the mayor's office which drew up lists of "rebels" and it was the mayor's agents—such as Gh. Şerban's own son in Nucşoara—who fixed the state delivery quotas, the main instrument of repression. They enjoyed a great deal of freedom in this, and the consequent arbitrariness and individual abuses made the already legalized terror even worse.

The main outcome of the repression was the gradual erosion of the peasants' autonomy to use the produce of their own land in the first place, and secondly to sell it or farm it out. Once all freedom in the use of their land had been taken away, with the massive support of the organs of repression, collectivization was practically achieved. The repression of the peasantry succeeded once they had been left with no practical means of survival apart from joining the collective farms. "In 1946–1947, twenty liters of grain sold for one million and people were doing better, then the taxes went up, and the quotas went up, too, so people would leave the field only with their rakes on their back, and not a single ounce of grain." (Marinică Popescu, Scorniceşti)

The repression of the Hungarian Revolution in 1956 signaled to the peasants that the West was not going to intervene to alleviate their plight. The demonstrations of solidarity in Romania created thousands of new political prisoners. In Timişoara an entire ad hoc internment camp had to be created to accommodate the mass of arrested students. Officially, there were no decrees legally enforcing collectivization, because the propaganda campaign at the time required it to be voluntary. Therefore, legally speaking, the peasants' lands, unlike the large estates, were not nationalized, but transferred "voluntarily" by means of contracts. In fact, however, the best land was confiscated and amalgamated in order to ensure the success of the kolkhoz (GAC). In exchange for the lands taken, rebellious landholders were given plots elsewhere. This contributed considerably to the subsequent post-1989 chaos, when properties had to be accounted for. The peasants also contributed to the confusion by declaring less land than they actually had in order to evade taxes. These declarations, recorded in the Land Register, formed the basis for establishing ownership rights after 1990.

The strategies of subjection

Legal terror and violent repression were the extreme forms of a complex process which resulted in depriving the peasants of all autonomy. The Romanian Workers' Party's policies vis-à-vis the peasants would not have succeeded had such policies not denied people control over the most mundane acts of their everyday life. Successful repressive policies are not necessarily guided by reason, and arbitrariness may in fact be more productive because an arbitrary policy is not predictable. Everyone should be made to feel that they can be hit any time. That was why, drawing on the Soviet model, the party encouraged denunciation and surveillance, although it was to be expected that people would use this system to settle personal scores by denouncing enemies, for instance, or those with whom they were engaged in litigation. The party was fully aware of this, but ultimately it did not matter whether those thus denounced were enemies of the regime and the denouncers its pillars: What mattered was the progress of the destruction of the bases for mutual trust and spontaneous collective action. The very efficient village support networks of the partisans were particularly targeted. As peasants had shared interests opposing them to the new regime, in places such as Nucşoara the party had to bring in outsiders and install them undercover in locals' houses in order to obtain any information. Things were easier in places with enough marginals likely to be recruited against the rest of the village.

Tales from Scorniceşti suggest the arbitrariness of communist power, which targeted even individuals who tried to find an accommodation with the regime. Unlike areas in Argeşul de Sus, in Scorniceşti there were no border guards, weapons, and conveniently located forests, and resistance to the new regime was therefore limited. But the terror was as arbitrary and comprehensive as everywhere else. "Father had his own house, and I lived separately. He had two bulls and a plough and, before he joined the collective, he said: 'I'm old now, I need a cow for milking. I'm going to sell this bull and buy a cow instead. One of his bulls was seized the next day, and he sold the remaining one and bought a cow. He also had a wooden shed for the grain; it was rather small, but they were after him, so they told him they needed that shed. When he heard this, father had a heart attack and died the next day." (M. Popescu)

The party's policies aimed at subjecting the peasantry as an entire class, irrespective of differences between individual peasants. Official documents attacked chiefly the *kulaki*, who could afford to hire labor to farm their lands, but this was only a stratagem. In practice, poorer peasants were equally harassed and hit, especially if they were hard-working and thrifty and so had a higher potential for autonomous action. As recognized even in top-level party debates, survival strategies often had unlikely effects. Many party members from villages, some of whom were called up for the army, and others who had menial state jobs, were as reluctant as other peasants to join the collectives. They were the people who, during a temporary liberalization of the regime in the mid-1950s, were among the first to leave the voluntary associations. Conversely, some of the *kulaki*, fearing arrest, themselves became agents of collectivization. The Cârstea family from Scorniceşti was a case in point. With survival skills built up during periods of hardship, they were shrewd enough to beat the communists at their own game.

When the collective farm was being set up, Cârstea, appointed by the communists, had been mayor, since 1947. But the Securitate caught him with his brother as they were putting up posters for the agrarian party. They caught them and took them "away." Even before they got "there," however, they had become communists. Elena, their sister, was the first to join the collective. She and her family, and his brother. And we all had to do whatever they said, otherwise they reported us. The woman is out of her mind, even now [in 1990, *author's note*]. When the new joint farm was created, she took the chairman to court. She beat him up, she went for his throat. We told her the old kolkhoz was finished, we are taking our land back. But she went on about it, how they had worked so hard. Yeah, they were not guilty and had done nothing! They stole in broad daylight, that's what they did! The other people worked.

But tell me, how did the collectivization start? Did the Cârstea family bring land into the collective farm?

No, they didn't bring anything.

So what did they do to join the collective farm?

Nothing, they just blagged their way in! They left their lands with the parents. And the others, who joined, being no better, had the other brother, Marin, as an accountant at the collective. He was Bărbulescu's trusted man.

And what happened to their parents' land? Did they continue to farm it?

Yes, they did, until 1958, when we all had to join the collective farm; there was no escape. We went in, with everything we had, all our assets.

And why did they have Florea as a mayor? Was he a capable man?

Ha, capable! They had nobody else, 'cos nobody wanted the job.

Had he been a Communist Party member for some time?

Nah! I've just told you how they joined the party! They were caught, and first they joined the party, then they joined the collective farm and then they made him mayor.

So, in 1946 Cârstea—Florea, that is—became mayor. So who started the collectivization? Florea, with Marin, and Elena Cârstea's husband?

They were a small group, who started some small-scale collectivization in the beginning, as an example to the people.

So what did this small-scale collectivization consist of?

Well, just a group of people, and some of the elderly, who joined with their land, because they couldn't work it anymore. Others were forced. They worked on the collective farm for eight years without the rest of us. And they had a very hostile policy towards us.

You mean, they put pressure on you?

Sure enough! My wife, on Palm Sunday, sees this group in our courtyard. "Where are you from?" "From district branch. You must join the collective." "Go away, mister, out of my courtyard! My husband is away, we don't have land, and I'll join when I think fit!" We had only been married one year, since March 1948. This happened in March 1949. And they started measuring the place up with a measuring chain, left and right. And my wife seized the chain, so they cut her hand off using the chain, and dragged it around the yard. Finally we had to go and see Dej!

So you went to see Dej. And what happened at the meeting?

I never thought I'd ever get to meet Dej! But a relative of mine worked for the Securitate, as a lieutenant, here in Slatina. And he said to me: "So, what are you going to do? They did you a lot of harm, these guys; you are now on file, they could have you arrested. They wrote in your file that you are against them." And after a while, indeed, they threw me out of the party. They called me to a party meeting, they asked for my card; this was in 1953. And there was one, Manoilă by name, he is no longer with us, and he tells me I'm excluded from the party. Others, also party members, spoke up and asked what I had done, for they were afraid that sooner or later it would be their turn. There had been the deviationism of Vasile Luca and Ana Pauker [...] well, the purges within the party [...] And they excluded me, too. I met with my relative, who advised me to write a memorandum to Dej. I wrote it up in Slatina, in a restaurant. I didn't think it would get there, but it was forwarded by one Chilea, an activist at the secret documents section at Olt district branch. He sent it at the same time as the secret documents, and it arrived at its destination.

> After two weeks, a car from Bucharest draws up, and this Olaru, the chairman of the party collegiums, steps out. He came in, asked what's up, why had I been excluded? I told him I had no idea why I had been excluded. He had my file, but didn't tell me. He called me for an audience with Dej. I was afraid they might arrest me. I told him I had no money to travel, but he took his wallet out and gave me the money. An appointment had already been made. I went over, I went in. Dej asked me if I smoked, offered me a cigarette from his own cigarette case. He asked me: "How are things in Scorniceşti?" I said "well." "Does Ceauşescu ever come?" "Yes," I said, "he comes sometimes." He asked what happened to me and I told him I had been excluded from the party. Why? "That's what I'd like to know too [...] fellow-members protested, yet I've been thrown out." And then he goes on: "I'll tell you why you were thrown out. Look, I've got your file here. You were married to the daughter of a kulak, one Vasile Câmpeanu." And I say: "Comrade Secretary-General, I'll be shot dead if I have a father-in-law, my father-in-law is dead. Vasile Câmpeanu is indeed a kulak in the village, but he's not my father-in-law. They made this up, to have something against me." A neighbor of mine had apparently made a statement. My mother and he had never seen eye to eye, and so he had made a statement. I explained to Dej why I had not joined the collective farm. I had no land and couldn't walk the seven kilometers to where the collective farm was in order to work. And, I said, "I'll join when the others in my village join too." And he said I'd done the right thing, conditions were not yet optimal and us folks should join when we thought fit. And he said we should be brave, 'cos these people lash out even against party members. He told me to go back home and carry on working, and that I shouldn't fear anything.

Nicolae Popa was among the lucky few. As a poor peasant, he found a job as a statistician at the village hall, and was among the first to join the Communist Party. He knew the Ceauşescu family very well. He shared a bed in the school dormitory with the brother of Nicolae Ceauşescu, also named Nicolae—because they had the same godfather—but who was commonly known as Nicu. Popa had a cousin in the Securitate, who facilitated the dispatch of his memorandum at the same time as the official party papers, so it was duly forwarded. Thousands of peasants in the same circumstances, people like C. Stancia's family in Constantineşti for instance, risked prison. This is what happened to them, as recounted by Stancia and his wife:

> In 1947—we were young lads at the time, aged fourteen, fifteen, sixteen—we decided to form an association to destroy the communist posters and stop people from joining the collective farm. In 1951, some twenty families joined, and then nobody else; they didn't want to join. Those who went in were the folks with little land. They were poorer, and could keep the remaining land. I don't think it went very well, but they said it did, and why aren't others joining. But we didn't want to join, and had formed our own association. We had hardly started anything when they arrested us. They took me to the Securitate in Piteşti. They came by night and took two of us each time. (The husband)

> One evening, a whole regiment of military showed up in front of the house; they were militiamen. And I say to my husband: "Cornel, wake up, the courtyard is full of strangers." They came in and started to search the house. "What are you after, mister?" I say. "Weapons." "Why, we don't have weapons." "Shut up and get some food ready and a change of clothes, for he's coming with us." "For how long?" I ask. "Well, one or two days, no more," the man says. And they took him away and he was gone for the best part of one year. (The wife)

> They arrested me in 1959, for political reasons. I was twenty-eight at the time. And we had this three-year-old daughter. They accused us of running a subversive organization, "River Olt Speaks." They confiscated everything, we were left with nothing at home. And yet people did not join the collective farm. And so they looked for ways to make people join up. They would turn up at night, and would take two or three of us each time and bully us. They took away the village priest, a decent, peaceful man, and a few others who had been members of the Iron Guard.[17] We had nothing to do with them, there were just about six of us fellows in the organization: I was one, Nicolescu Ştefan, Marinescu, and a few others. We had set up this association after the communists came to power. They sentenced me to eighteen years' hard labor. But first they beat me up brutally. I was kept for six months at the Securitate. We told them what they wanted us to say. We said just write down whatever you want, mister, we're going to sign. I was given eighteen years for belonging to a subversive organization which undermined the socialist regime. My lawyer was unable to help. Later, after Piteşti, there was another trial in Bucharest, and

> I tried to explain that I hadn't done anything, that I confessed because they had beaten me up, but the sentence was upheld. I served it in the Piteşti prison. From 1959 to 1960, for they freed us after one year. So six months in Piteşti and seven in prison. Nicolescu was there, too, and Marcu Marcel, also sentenced. In prison they beat me up once more. There was this large room with bunk beds on top of each other, there were many of us. There was Marcu Marcel, Liță Marin, Dumitru [...] (The husband)

> I appealed, I went around all the neighboring villages saying they were good lads, they hadn't done anything wrong, and I tried to have them pardoned. Then I joined the collective farm, but they wouldn't have my land, 'cos they said we were enemies of the people. I had to beg them. After that, the entire village was collectivized. Three days after the lads were taken, the whole village was collectivized, they queued up to join they were so afraid. We were first told that they would take our land and give us another plot to work, just outside the collective. But then we were allowed to join too. (The wife)

> A year later, our sentence was lifted. For all of us. First they set us free, then there was another trial ten years later, for our "rehabilitation." In 1969, we were taken to Bucharest again and we were tried and arrested without evidence. The entire village had joined the collective in the meantime; everybody was in. (The husband)

From the opposite camp, the party activist Gheorghe Şerban from Nucşoara tells the story of collectivization from the perspective of the party cadres who supervised it. Not in Nucşoara, however, where "officially there had been no attempt at an association; something was tried in '54–'55, but the land was not suitable for farming. We set up a collective farm in Slatina instead, but it didn't work because the land was no good." The eighty hectares confiscated from the political prisoners had been swapped for lands in Slatina, which were then supposed to be "combined" as a start-up for the collective farm. Collectivization was completed, however, in Domneşti and in the lower areas of Argeş country. Constantin Paşol, the secretary of the local party branch in Nucşoara, was there, but he openly admits that it did not work. Although even today he is proud to be a communist, he believes that it was the communists' worst mistake to coerce peasants into the collectives. Gheorghe Şerban, too, admits that collectivization was "slow." Although he declined to go into detail, he is proud to remember that he and a few other cadres had been responsible for setting up the first communal farm in a village not far from the birthplace of the great reformer and promoter of the peasants' cause, Ion Mihalache, founder of the Peasants' Party, who himself died in the gulag.

> Then they gave me a job in Topoloveni and it was there that we formed the first collective farm. We went around looking for volunteers to join the collective. It wasn't easy because there were many *kulaki* there. We went around talking to people, but there were boyars, especially one named Păun. So we went around and said "look we're setting up a collective farm," and the party adviser was with us. It took me about a month at the rate of two or three applications per day to set up the first collective farm. There were ten *kulaki* there who had influence over the peasants. In the evening, it appears that the *kulaki* would call the peasants around and ask them: "What did the communists tell you?" people reported to us. And we told them "let's make an application and have a collective farm here." And it took us two months to gather the necessary land, and we had party cadres with us, who knew how to go about things, 'cos they had been to party school. We would talk to people, and they were not bad folks, not unfriendly; they wouldn't jump on us in the street at night to threaten us. We also went round to Mihalache's house, and so we made the first collective farm in Muscel county. Then it all spread out and others were set up, too, by experienced people sent by the Central Committee. Later I worked at the prefecture in Câmpulung for about three years, and we set up many collective farms there.

The voluntary association in Nucşoara failed not just for lack of adequate land—the plots were widely scattered—but also for lack of participation. The activists who joined up, such as Şerban Gheorghe and Constantin Paşol, and people such as the Cârstea family from Nucşoara, being rather conservative, did not place their land in the collective farm, opting instead to farm it themselves. "However, they placed the sheep in joint ownership. They put the sheep together and paid out their quotas out of the joint produce. Those who were not members had to pay hefty quotas instead. They had to deliver two lambs, two sheep each." The first voluntary associations, therefore, were seen as a means to evade the heavy delivery quotas burdening the other peasants, because such schemes as well as the kolkhoz benefited from considerable discounts. But who was in a position to put such strategies into practice? Only those who were above the law, the communists and the collaborators of the Securitate, the very people who preached the virtues of collectivization.

The Slatina "state agricultural enterprise" (IAS) was one of many such abortive projects, although at the time a lot of money was invested in it and the farm was placed on the "protocol circuit" and visited by ambassadors and the like, as one former director fondly reminisces. Conversely, the Scorniceşti "agricultural production cooperative" (CAP) was a great success, and the only person who refused to join, Nana Iana, was the first to have her house demolished when systematization was launched. In 1989, the cooperative "had liquidity worth 300 million lei, produced 5,000 tons of poultry meat, 17,000 hectoliters of milk, 160 tons of beef, 160 tons of pork, and had a 16-hectare hothouse the produce of which was exported to Western Europe. It also had 3,000 hectares of arable land. It was the largest in the country, both economically and in terms of assets," says Ion Neacşu, former mayor. However, large exports did not score very highly in the households of the Scorniceşti locals, although, comparatively, cooperative members there were privileged. "The collective with the highest dividends paid to members in Romania was here in Scorniceşti. They used fertilizers, so production went well. It produced enough for the state and for the people. But they received just enough to make a living, two or three kilograms of grain per working day. Perhaps in some families only one member was active, and there might be six mouths to feed," Marinică Popa explains. "Mother" Stancia is even more emphatic in her condemnation of the cooperative: "It wasn't good in the beginning. They'd give us between two and a half and three kilos of grain per day; it wasn't enough to live on. For when we went we'd score seventy to eighty points per day, and when we could we'd go for a whole day, especially when harvesting, but one could rarely score 100, and one wasn't able to live on less. Women had to do 150 days per year, and men 200. We couldn't do so many days, especially us women; we had children, pigs, and poultry to look after; I couldn't go daily. They'd pay us two or three lei per working day. What was one to do with two or three lei and two kilos of grain?"

If the peasants survived at all, it was by taking up alternative jobs in the city, by joining the large-scale exodus from the countryside, and by using the produce from the gardens they had been able to keep contractually upon joining the state farms. The confiscation of these gardens at the end of the Ceauşescu regime was a disaster for the peasants, but also one for the regime, which at that stage, quite clearly, lacked the means to enforce the large-scale implementation of such an absurd and unpopular decision.

> In 1987 the persecution of the collective farm members, members resumed. They decided that we should no longer have plots for our personal use. When we joined the collective, we made a contract—I turn so much land over to the collective and keep the rest, let's say thirty acres. And then suddenly they got this idea in Bucharest, and put it into Ceauşescu's head, and he signed, to the effect that we were no longer entitled to even one single yard. "We wish to inform you that, dated 21 March 1987—to the attention of Mr Popa Nicolae—we wish to inform you that as of 1987 you are no longer entitled to work the 0.36 hectare plot, representing arable land, which is going to be incorporated into our unit. We wish to draw your attention to the fact that unlawful activity on this land will be prosecuted." Signed by the then president. I don't remember, I think the name was Turcin. And so they came and they [...] I had only the house left [...] they came and they sowed barley and corn [...] they said this was going to increase the output, so we'll pay Romania's external debt [...] from our gardens.

This move destroyed what was left of the regime's popular appeal, even in Scorniceşti, the place with the best food supplies in the country, which was declared a town in 1989. The peasants had enough of Lina Bărbulescu and of Turcin, the collective's chairman. Lică Bărbulescu enjoyed some popularity, but no longer. There was even a satirical song which went the round of villages:

Lică surely it's a sin
What you're doing with Turcin.
You've laid waste our lives and lands
You take the bread from our hands.
You uproot our vines and trees
And bring us to our knees.
Let crows and worms pick at your eyes
And your bones turn to dust as they rise

To be scattered on winds for the blames
Which attend your cursed names.
And let Vlad the Impaler strike
And hoist your skulls on a spike.

They were not to know that in just a few months they would be looking back with nostalgia upon those times. CAP Scorniceşti, the largest collective farm in Romania, collapsed on 22 December 1989, alongside the communist regime, leaving the export orders for the produce of the huge greenhouse unfulfilled. People have mixed feelings regarding this event, but the large-scale plunder which followed left everybody in a state of shock. For some, it was like the sinking of the Titanic.

> Scorniceşti produced all that was needed; tinned food, meats, soft drinks—Pepsi was made and bottled here. And the greenhouses went full blast, I can tell you! People would work there full-time; they no longer bothered with farming, they were employed full-time in small manufacturing. They didn't need grain to work, and of the gardens almost nothing was left; people barely had a thousand square metres each. If they needed anything, they would go out and steal, that's the truth. For harvesting, they had the schoolchildren who were doing mandatory work, and whatever was left when they finished in the fields people would just take to "top up" their supplies. Effectively, when they had Ceauşescu killed in 1989, I warned: "Well, sir, if the Scorniceşti collective farm goes bankrupt, our whole economy will collapse." And this is exactly what happened. (Focus group intellectuals, Scorniceşti)

Twelve years later, the assets of CAP Scorniceşti had still not been disposed of. But by then, that heritage was worthless: The fortune which the Ceauşescus had tried to amass for their clan, like all the projects of the socialist regime, had come to nothing.

Landscape after battle

Anyone walking down the high street of Scorniceşti—formerly the village's main road—on the morning of the last Monday of June 2001 would have been treated to quite a sight. A few men armed with metal bars were hitting desultorily at some unfinished concrete walls, dislodging a fragment now and then. They tried to dodge a few bystanders, an older man, a younger man, and a woman with an apron, who were all shouting something. A neighbor in a denim shirt and an old woman called Nana Veta, who knows everything that happens in the village but is invariably told to shut up whenever she tries to speak, stood staring at the mess. An ageing man in a white shirt and spectacles, looking like someone from the city, was standing in the middle shouting encouragement to the demolition agents. He was a former biographer of Ceauşescu, and that was his father's land, legally reacquired after eight court cases. He was paying the demolition team from his own pocket. The unfinished building being demolished was supposed to be the stables which the neighbors who lost the lawsuit were attempting to build. Made from reinforced concrete, the structure was typical of postcommunist rural architecture. The structure shook and twisted as the metal bars knocked the concrete flesh off its iron skeleton. A man in a bright red shirt was reigning over this chaos. He climbs up onto the ruins holding a measuring tape, and stands among the frightened, howling stray dogs, shouting above the bedlam. His words are reasonable: "You don't have the authority. Sir, please, if you don't mind, you don't have the authority. I am fully authorized." And he is right. This man is the legal owner.

Twelve years after the fall of communism such scenes became typical. Difficult as it had been to amalgamate the land, it was even more difficult to subsequently subdivide it. The two opposing parties were both in the right. The owners of the stables were right because the Land Register listed the plot on which they were building not as the biographer's property, but as the property of Nana Veta, their aunt, who had sold it to them. They had a document from the Prosecutor's office showing that the 1954 Land Register had been tampered with after 1990. But the authorities had not identified the forger, and the biographer insisted that he was not responsible for the tampering. He, too, was in the right, especially as his claim in court had not been countered, and the court had found in his favor. His opponents, being people with little education and little money, had failed to produce witnesses—i.e., Nana Veta and the man in the denim shirt—at the Bucharest tribunals. The local authorities were on their side, especially as they feared this could open up a Pandora's box. The house of the stable owners stood on the biographer's family land, even though nobody could tell for sure if it had belonged to his father or to his aunt. However, like everyone else in town, the stable owners had had their land expropriated during systematization. The new residents in the housing blocks and villas were given homes, but not the land they stood on, which had remained in public ownership. When the first land restitution law came into force in 1991, almost no-one in Scorniceşti thought of applying for ownership of the land under their home. The next law (2000), more comprehensively anti-communist in its thrust, attempted a return to the situation of 1945, just after the agrarian reforms. This time, the local authorities rose to the challenge and told the new owners to reclaim their lands. But in a few cases it was too late: The plots had already been claimed by the older owners, whose lands had been abusively expropriated by the communists. The court cases were dragging on, however, because in Romania such litigation takes four to five years on average. In one case, the tribunal found in favor of the former owner. He won the land under the home of someone who had lost their own home and land during systematization. This created an intractable situation, in which both sides were in the right, but both were victims.

There are tens, possibly hundreds, of thousands of similar cases. Over one million Romanians are still struggling in the courts because of the legal chaos created over the restitution of land. A decree in 1990 by the then acting president, Ion Iliescu—who was subsequently elected for two terms, largely on the back of rural votes—returned small private plots to their peasant owners. These were the gardens expropriated by the communist regime just before its demise. It was a gesture which earned the new regime considerable support from the elderly population, who depended on these gardens to feed their poultry and livestock. New laws passed in 1991 were more ambitious. The aim was to reconstruct rural property and to take the first step towards creating a land market. Law 18 of 19 February 1991 aimed explicitly at the "reconstitution or construction of property." It was not, therefore, simply a law covering the return of property, but also its privatization. This was new territory: the term "construction of property" was a legal and linguistic novelty, which appeared more at home in the discourse of communist social engineering than in the customary vocabulary of liberal democracy. Privatization normally refers to the transformation of state property into private property, but the lands of the collective farms were not state property, they were the peasants' private property. Law 18 in fact allocated lands to people who were not their rightful owners. Article 8 stipulated that: "The present law benefits members of collective farms who either voluntarily contributed lands or whose lands were taken over by the collective, as well as, under civil law, their heirs, cooperative members who contributed no land, and *other specifically nominated persons*. Title deeds can be issued on demand to incumbents who wish to have their ownership rights guaranteed, as per the present law, over surfaces of at least 0.5 hectares per person and not exceeding ten hectares per family, in arable equivalent."

Law 18 was the most eloquent expression of the ambiguity of Romania's postcommunist regime. President Ion Iliescu was a former Communist Party First Secretary: Given the particular circumstances in Romania, the Soviet Bloc's most communized country, in May 1990 Iliescu was elected with eighty-three per cent of the vote in the country's first free elections. His regime had a strong support base in the army and the Securitate, whose contribution had been essential to Ceaușescu's downfall.[18] Caught between the requirement to follow the same path as the other countries of the former Soviet Bloc towards democracy and a market economy on the one hand, and the demands of the particular circumstances that characterized Romania on the other, Ion Iliescu's regime chose to do things differently compared with the new regimes installed in Prague, Warsaw, and Budapest. The regime's most defining moment was Law 18, which attempted to please everybody. Those who had been coerced into joining the collectives received a few hectares, a maximum of ten on a first-come-first-served basis, and not necessarily their own former plot, but its "arable equivalent." But Law 18 also returned land to cooperative members who had not actually owned land and who, therefore, had not received anything in the 1945 reforms. These were mostly people who had been resettled in the countryside to support and speed up collectivization—activists such as Gheorghe Șerban's comrades. Lastly, the reference to "*other specifically nominated persons*" in Law 18 seemed designed to open up the floodgates of arbitrary claims. Thus formulated, the law gave the commissions which applied it locally, and which comprised the mayor and his associates, total freedom in nominating recipients. In all regions, the last two categories of recipient obligatorily included the intellectuals resettled in the countryside by the communist regime as a replacement for the village schoolteachers and priests who had perished in the political prisons. These people were not locals. Some, but not all, had become landless members of the collectives, and had not contributed land to the communal holding. However, the postcommunist state deemed it unfair to leave them without property.

The manner in which the land redistribution commissions applied the law fully exploited its potential for disaster. This is how the mayor of Scorniceşti, elected no earlier than 2000, innocently put it: "Ultimately, it was up to the mayor, for the land stayed the same, it was not wider, and it was not smaller." In effect, the land extended or shrank according to the will of the mayors and the commissions. After all, the mayors who applied the law had not been elected. They were directly appointed by the Iliescu regime. Some had grabbed the job in the scramble following the collapse of Ceauşescu's regime in December 1989, but by February 1991 the only mayors left in place were those confirmed by the new regime. There was still one year to go until the first local elections. People such as Ceauşescu's biographer had been members of the Land Commission, with full access to the Land Register, and they would have had plenty of opportunities to tamper with the records.

The Commission distributed the land starting from a few basic rules, some of which took the form of political signals from the very top, i.e. from President Iliescu himself, while others derived from the petty interests of commission members. The president made public his dislike of excessive land fragmentation, which historically had been the major problem affecting Romanian farming. He spoke about the positive heritage of communism, such as the pools of farming machinery and the irrigation systems. Ideally, he believed, the newly endowed peasants should have registered their lands with free farming associations on the French model—which he specifically cited—or the Danish model, the old template of the Peasant Party. At the time, however, the reinvented Peasant Party, harking back to the realities of 1945 and believing that the peasants were all wanting their lands returned, was totally opposed to the collective farms. But the peasants had stopped farming the lands years before, and they were not keen to embrace legislation which might have entailed an obligation to actually work the land. "Who's going to work the land? We're old;

the children have all gone to town." Selling land was almost out of the question. Foreigners and Romanians residing abroad were banned from buying it. If inheriting land, the latter were required by law to sell it within one year. As far as the sale of unincorporated lands was concerned, there were several categories of claimant who had pre-emptive rights: joint owners, if such existed, followed by any neighbors who might have wished to supplement their own land, and, lastly, the Agency for Rural Development (Agenţia pentru Dezvoltare şi Amenajare Rurală), a short-lived state institution. If none of these expressed an interest, and if the Agency did not offer to buy it on behalf of the state, "the land was to be sold on the open market" (Article 48). In the circumstances, no land market was ever created. Owners who had had their land confiscated by state farms received shares in the state farm (IAS).

The Land Commission became the chief actor. The 1991 land law, which seemed to draw its inspiration from party documents from 1948–1958, suggested that morally only individuals willing to work their entire land were entitled to it. This favored the creation of another round of voluntary associations and went against the principle of returning lands on the original sites. Whoever wished to withdraw from the association would receive a plot "on the margin." In Scorniceşti, this became the typical pattern. The tools inventory was not returned to owners, but in the aftermath of the Revolution of 1989 everybody had grabbed what they could, except the machinery. The *specifically nominated persons* turned out in practice to be the mayor and his clients, especially his family. This was nothing new. In Nucşoara, an uncollectivized area, Church and school properties had been expropriated by decree in the same way as in the collectivized regions. But as lands could not be left untended, the properties were taken over by the mayor, his acquaintances, and his Communist Party comrades, those arch-enemies of private property. Gheorghe Şerban, who before the war had had less than three acres, saw his assets increase fourfold after these confiscations, so much so that he could endow his children too. The great land restitution of 1991 had already had a dress rehearsal in Nucşoara a long time before, and had seen Communist Party apparatchiks turn surreptitiously into landowners.

In fact, the great land restitution turned out to be only a sideshow. Aside from the mayor's clientele, there was no rush to issue title deeds to anyone. Consequently, there were no sales of land to speak of, even when the famous rights of pre-emption had been observed. By 1995 not even a quarter of title deeds had been issued. By 1996 some 600,000 lawsuits were pending, because the new owners received land on the property of other people who had not been allocated the ten hectares that were due to them. With the conflicts over partition and inheritance, forgeries and imprecisions in the Land Register, the reluctance of state archives to issue the original title deeds, the picture became very complex indeed. In the circumstances, the peasants relied on the local authorities, much the same as before. Whether or not they became landowners depended on the benevolent attitude of the Land Registry man, the archivist, the Land Commission, the mayor and, ultimately, the county prefecture which issued the title deeds. By 2002, optimistic estimates placed the percentage of returned title deeds at around sixty—although in many instances the land itself had been returned—and the duration of court cases at six years. But informal sales, without the paperwork, were a common occurrence, and building on sites of uncertain legal status was the rule rather than the exception.

In 1996, the former communists lost the elections. The new winning coalition represented a wide spectrum of interests. The largest party in the coalition was the National Peasants' Party, founded by Ion Mihalache and Iuliu Maniu. A new law was passed in 1997 which was intended to correct Law 18. Law 169 of 1997, known as the "Lupu Law," after the name of its main proponent, the Peasant deputy Vasile Lupu, allowed for the return of plots in excess of ten hectares within the limits of the available land, and for the restitution of lands from state farms. The state farms had been protected by the former communists, represented in the coalition of 1996–2000 by the Democratic MP Triță Făniță. His opposition to the privatization of the farms, which had made headlines in *Business Central Europe* in 1997, was effective enough to obstruct the application of the Lupu Law. It took around four years for a law to be passed on the privatization of state farms. Wherever there were no state farms, the law was applied by the same mayoral offices, which led to new waves of lawsuits.

Ten years after Ceauşescu's downfall, the allocation of plots in the countryside on the basis of temporary ownership certificates—pending the issuing of permanent title deeds—showed considerable success in recreating small properties under 5 hectares. Having increased from 75 per cent in 1918 to 76 per cent in 1949, small properties accounted for 81.6 per cent of total arable land in 1999. Plots between 5 and 20 hectares, of which there were never many, dropped from 23 per cent to around 18 per cent, while bigger plots—i.e. farms proper—were practically non-existent. The fragmentation was initially offset by the association schemes recommended by President Iliescu, but by the end of the 1990s most of them went bankrupt.

Table 2. The size of rural properties

Hectares	Distribution by size (%)		
	1918	1949	1999
Under 5 ha	75	76.1	81.6
5–10 ha	17.07	17.8	15.1
10–20 ha	5.49	4.89	3.1
Over 20 ha	2.54	1.20	0.2
Total available land	3,280,000	3,067,000	3,211,507

Source: National Statistics Office (INS)

In Scorniceşti, the assets of the large cooperative were initially divided into three associations. The largest of them, Comagri, came quickly under the control of former comrade Turcin and his people. It had the largest membership and, with 2,000 hectares, it was also the most extensive. In the neighboring villages, which had been coerced into large-scale "agricultural production cooperatives" (CAPs), small-scale local associations branched out one by one. They took with them only the land, while the other assets of the former CAP remained within Comagri. Only once the amicable partitions failed, a few years later, did claimants resort to the legal system. But what was there left to claim? Like everybody in the country in the immediate aftermath of the 1989 Revolution, the peasants of Scorniceşti had behaved like their literary counterparts in the stories by Ilf and Petrov written after the Bolshevik Revolution. Everyone grabbed and took home what they could. The irrigation system was dismantled and pieces of pipe were used as drains in gardens. Whatever was found in the head offices of the cooperatives—the former boyars' manors—was stolen. Livestock was stolen too. The large stables in Constantineşti are now empty; rain pours through holes in the roof. The slaughter-house is empty too: above the entrance a couple of forlorn animals' horns are a sinister reminder of the building's past. The greenhouse which used to set export records was literally smashed. "Scorniceşti is like Hiroshima," says one of the boys showing us around. Fish swim in basins in the flooded hothouse, now without its glass panels. One wonders where they came from. Stray dogs sleep in the greenhouse, and a homeless man sleeps in the slaughter-house. For most of the transition period, large sections of land were left unworked and are now swamped by weeds. People blame each other. At least the greenhouse was the object of a police investigation. It was dismantled and sold off piece by piece. The culprits were never found. The more enterprising individuals with more land formed small-scale family associations and fought to have their ownership of the land recognized. They are doing well: Marinică Popescu's son, as well as the Stancia family, can now sell their surplus in the town market, and both of them have been able to buy a tractor. They were lucky to live in villages away from the commune centre: In Scorniceşti itself they would have been allocated plots on the margins.

Larger associations had a more uncertain fate. When we visited Comagri in 2002, it was under investigation. The villagers hired the services of a son of the village, who, being privileged enough to be born in Scorniceşti, was able to build a career in the Finance Department of the Ministry of Internal Affairs. This old Securitate employee wrestled with mountains of paperwork left behind by Comagri's management. The bankruptcy was so comprehensive that the land had to be farmed out to a Bucharest firm with good political connections. The same party, which also owned a farming association, managed to have a few local councilors elected. It was a self-declared party "of the workers," positioned to the left of the postcommunist Social Democrats, the true heirs of Ceauşescu. The Constantineşti branch went into bankruptcy, and the picturesque manor of the Coculescu family, the farm's head office, had its electricity supply cut off because of arrears in paying the bills. The management of Comagri was a disaster, and the peasants received no dividends from the association which had incorporated their lands. They were even asked to pay money to cover the association's outstanding debts, and the slaughterhouse was seized by debtors for the same purpose. When people complained that it had been undersold, they were asked to settle the loans and the debts out of their own pocket.

The association formed in Mogoşeşti, on a site of the former landowner Dr Voiculescu and run by the son of the doctor's former butler, was doing much better. Members called it the "Dr Voiculescu" Association, for luck. In his old age, the former butler became an alcoholic, but his son, an engineer, was among a handful of communist-trained agronomists with confirmed managerial talent who had not dipped into the association's cash pot and had not bankrupted it. In 1992 the association had only two Romanian-made harvesting machines, costing less than 5,000 euros. Its yield in wheat that summer amounted to double that in the inflated national currency. That was a good year, and they were able to take out a loan from the World Bank. They had difficulty paying it back, as the value of the national currency dropped, but with the loan of two million Deutschmarks they bought new machinery, leading to a substantial increase in the firm's output, and so they coped.

Furnica, the smaller of Scorniceşti's three farm associations, had more problems, but they were on their feet too. The engineers in management looked back nostalgically to the heyday of the former cooperative, CAP Scorniceşti, and tried to find some international conspiracy to blame for its demise. They needed the association to survive because they themselves had brought no land into the collective. Being graduates of agricultural colleges created during communism, they had been among the "*other specifically nominated persons*" brought in late by the regime, but not all of them had been given land under Law 18, so they needed the association to make a living. The story of the Furnica Association deserves to be told:

Please tell me, how privatization was done after 1989 and how your own association was formed?

After 1989, three enterprises were formed out of the former cooperative farm (CAP): Comagri, Furnica, and Frăția. To start with, we had some ninety hectares, and by 1992 there were about 450 members. Subsequently, some people opted out, leased their land to someone else, and there were about 385 people left.

What did you have at that stage in your patrimony?

Just land.

And how did you go about acquiring equipment?

Equipment? We bought a few tractors and a harvesting machine and later we had to go to court over the former farm patrimony belonging to our members. After the case, four years later, we managed to get something, far less than would have been due seeing what they had brought into the collective farm fifty years ago or what their due share from the Scorniceşti Farm was.

And when the former cooperative farm (CAP) was dismantled, what was people's reaction? Were they pleased?

They were, of course. They went "right, we're rid of serfdom, let's privatize," and then they all started to steal.

They hadn't been stealing before?

Of course they had!

What about the police, was there no police?

Well, there was, but you know the people in Oltenia: If the mayor steals, and the party secretary steals, then they all steal.

And are they still stealing today?

Well, they would, if they could.

How's that?

Today it's the law of the jungle; it's the end of the road. Then [under communism, *author's note*], if you slapped someone, they would take you to court; but now it's every man's private land and they will defend it. If they catch you stealing, they take you down to the police station, and they treat you much worse there nowadays, or they beat you up to a pulp, 'cos it's less of a hassle.

Is it true that you had a very large output under communism?

Oh, yes, it's very true.

So what has changed in the meantime?

Here in Scorniceşti they would do crop rotations every two years, rather than every four years. Today we don't have enough fertilizer. We experimented with a field of sunflowers, where we used chicken pooh; you should've seen the difference! We really used to have the technology here! The soil is now depleted, and what are we doing about it? Nothing! There's not enough capital.

In my understanding, some considerable revenue from the sale of the cooperative was squandered.

The money was given to the people. Everyone received something. All proceeds were divided then, after the sale.

But why did you have to go to court to divide the assets?

Because we couldn't agree among ourselves, we had to resort to the law. Everybody wanted to split, but when we got down to it, nobody wanted to share: You can't have this, you can't have that. So I said, here's our group, and there's yours, your people. We could've had a holding here and keep it as a farm. There was this bit of real countryside left, and they had to destroy it.

How many members have you got now?

Some 120 families.

And how is labor organized?

We all work the land and each family gets a quota of the produce. Last year we gave them 250 kilograms per hectare. This year, we're hoping to reach 500 kilograms per hectare.

I understand that here in Scorniceşti you have family associations as well. Why might some people prefer this form of association rather than a company such as the one you're in?

Well, this is up to individual people. There were other things too; those family companies didn't really pay taxes immediately after the

> Revolution, while we in the bigger companies had to pay state taxes, fees, this and that.
>
> Now it's no longer the case, because everyone wants subsidies, and when you apply for them they ask you for proof you paid.

Next to Nucşoara, in Domneşti, the largest commune in the area and the birthplace of Elisabeta Rizea, the cooperative farm disappeared in a similar way, overnight. People simply took their orchards and their livestock back. Today, the motorway meanders through neatly partitioned plots, mown and harvested on time, where haystacks dry peacefully in the sun. Here, the cost of the farm's breakup was minimal and the peasants could resume their lives as though the fifty years of communism had been just a bad dream. In spite of the conflicts and the court cases, even the breakup of the farm itself was less traumatic than in Scorniceşti. The farm could not be saved. A special law was passed allowing the former political prisoners (peasants deported to the gulag) to take back their lands. Under the communist regime, these lands had been allocated to other villagers. Tha latter had had their own lands confiscated in 1945 and joined to Pavel Paul's estate to create the first state farm. When they lost their land in favor of the former political detainees, the peasants walked onto the lands of the state farm, carved them up, and helped themselves to new plots. The public farm management sued them over this, and they in turn sued the farm management, claiming their former right to that land. They also had legal disputes with the former prisoners, who, they claimed, should have been the ones to get land from the farm rather than getting their former lands back. Caught between all these entitled claims, the state farm quickly collapsed.

But this was a state farm in a mountainous area, which had done little beyond centralizing sheepherding, which until then had functioned independently in the mountains around Nucşoara. As herding had survived independently outside the state farm, the independent shepherds took over things smoothly when the farm collapsed. The situation of state farms in the plains was different however. Their labor force was the agrarian proletariat, recruited from the social rejects of communism. Ioan Gavrilă, a former partisan from the Făgăraş area, had been amnestied by the Securitate and sent to work on such a farm until his retirement in 1989. He was concerned for the fate of these day laborers in the event of the state farm being dismantled. They were people who had nothing in the whole world, "people without families, men who went into hiding because they were unable to pay alimony, single mothers, former and, most certainly, future offenders, Romanians, Magyars, Gypsies."[19] Communism left a legacy of an agrarian proletariat, of this dispossessed mass, after inheriting a country in which every peasant had owned a strip of land, even if it was scarcely large enough to allow the peasant to survive.

Notes

1 Alexandru Ioan Cuza (1820–1873), son of a lesser Moldavian nobleman, elected joint ruling prince of the Principalities of Moldavia and Wallachia in 1859, an election which led to their union. In a series of reforms, he expropriated lands owned by the so-called dedicated monasteries, whose revenues were used to support Greek Orthodox monasteries in the Ottoman-controlled Holy Places. In his land reform of 1864, two million hectares of land were distributed to 500,000 peasant families. He was ousted in 1866 in a *coup d'État* and died in exile in Heidelberg.

2 The peasant revolt of 1907 started in Moldavia and spread southwards into the Danubian plains, before being brutally suppressed by the army. At the heart of the conflict was the system of farm tenancy: Absentee landlords leased farmland to tenant farmers who, in turn, farmed out plots to the peasants, imposing increasingly harsh contracts.

3 The Peace of Adrianople (1829) marked the end of a Russo-Turkish war in which the Russian armies defeated the Ottomans. It eliminated the Ottoman Empire's monopoly over agriculture in the Ottoman-controlled Romanian Principalities and opened up the export of grain to Western Europe.

4 Roberts, *Romania*, 67; Mitranyi, *The Land and the Peasant*, 316.

5 General Gh. Fl. Petrescu, *Monografia județului Olt*, 60–5.

6 In World War II, under the Antonescu government Romania became Germany's ally from November 1940. On 23 August 1944, King Michael secured the support of the political parties, including the communists, arrested Antonescu, and, as a result, the Romanian army changed sides. Subsequently, the Romanian army actively participated in the expulsion of German soldiers from the country's northern regions and advanced with the Allied armies as far as Budapest and Prague.

7 Iancu et al., *Colectivizarea agriculturii în România.*

8 The Romanian gulag. The Danube-Black Sea Canal was a seventy-kilometer-long waterway set in an area of marshland and rocks. It became a concentration camp for some 50,000 prisoners before being dismantled in 1954.

9 See Deletant, "Studiu Introductiv," 32, in Oprea, *Banalitatea răului.*

10 Werth, "Introduction, Le pouvoir politique," 51.

11 ANIC, fonds C.C. al P.C.R. (Central Committee of the Romanian Communist Party); Cancelarie, file 59-1950, 65–82.

12 Decree no. 115-1958, published in *Buletinul Oficial*, vol. VIII, 19–30 March 1959.

13 The marginalization of Ana Pauker (1893–1960) and her group was effected in 1952 by the victorious faction of Gheorghe Gheorghiu-Dej (1901–1965). The conflict reflected the divide between the communist leaders,

such as Dej himself, who had spent years in prison, and those who, as refugees in the Soviet Union, had returned to Romania with the Red Army. In the 1930s, Ana Pauker was an important agent of the International and was active in France in circles around Maurice Thorez.

14 Gh. Gheorghiu-Dej, meeting of the Secretariat of the Central Committee of the PMR in 1950. In Cătănuş and Roske, *Colectivizarea agriculturii*, 184.

15 Ibid., 187.

16 Ibid., 194.

17 A right-wing Christian nationalist interwar organization. The Iron Guard was a grassroots movement, banned in 1941 by Antonescu after attempting a *coup* during which numerous people, mostly Jews, were assassinated.

18 Ion Iliescu, elected provisional president on 22 December 1989, was subsequently re-elected in 1990, 1992, and 2000. (He lost the 1996 elections.) He successively organized and reorganized the successor to the Communist Party, first as the National Salvation Front (FSN) in 1990 and lastly as the Social Democratic Party (PSD) after 2000.

19 Ogoranu, *Brazii se frîng*, 329.

CHAPTER 4

The Invention of Social Conflict

A failed project

"The rural poor – i.e. the proletarian and semi-proletarian agricultural workers – must be on full alert and permanently mobilized in the struggle against kulak exploitation. [...] The kulaks should be permanently under attack from the rural poor, and must be kept under pressure in order to diminish their moral, political, and economic power," reads the Report of the Political Bureau of the Central Committee of the Romanian Workers' Party (PMR), presented by Gheorghe Gheorghiu-Dej, the party's First Secretary and the protector of Nicolae Popa, the peasant from Scorniceşti.[1] Dej was simply explaining to Romanians a strategy inherited from Lenin via Stalin, who summed it up in the following terms: "The alliance between the proletariat and the peasantry is an alliance between the proletariat and the rural working classes. Such an alliance can endure only if the poorer peasants are organized into a support base of the working classes in the countryside. Therefore, in the current circumstances of the dictatorship of the proletariat, the alliance between workers and peasants can be achieved only in the spirit of Lenin's well-known slogan: Seek support among the poorer peasants, establish a strong alliance with the middle peasants, and never cease to struggle against the kulak class. Only through the application of this slogan can the masses of the peasantry be guided along the road of socialist construction."[2]

The project was clearly signposted, but its practical application was much less clear-cut. First of all, there was no clear definition of the kulak, and quite often the distinction between kulak and middle peasant was blurred. Following the Soviet model—the general template for the socialist transformation of agriculture—in communist Romania all those with the potential to cause trouble for the regime were categorized as kulaks. However, in Romania as well as in the Soviet Union, the peasants were more likely to admire the upwardly mobile, wealthy kulaks than the poor "proletarian" peasants. The kulaks were people who had worked their way up and were at the center of patronage networks which benefited the village, whereas the day laborers were looked down upon as little more than drunken idlers.[3] Boundaries were fluid, and the peasants preferred to identify themselves with the "kulaks". In Nucşoara, as few as ten families had land in excess of ten hectares; they included families such as Paul, Arnăuţoiu, and Şuţa. The average peasant had around three hectares. There was no evidence of a self-perceived class difference among categories of the Romanian peasantry, just as there was no evidence of it in Russia.[4] In fact, there existed no such distinction, because the peasants shared basic economic interests, namely low taxation and access to cheap, guaranteed state loans for the development of their farms. Lifestyles and land issues were the same everywhere, and family wealth varied from one generation to another, according to number of heirs, sons to marry off, and daughters to endow, so nobody was ever looked upon as an "exploiter". The Bolshevik definition of labor for hire was not suitable either, because sometimes sons from wealthier families would in fact be working days for poorer peasants, while people with tools would offer their services to anybody. Making life difficult for the kulaks was rapidly exhausted as a strategy, and it was equally hard to obtain the neutral indifference of the middle peasants and the support of the poorer peasants. The regime therefore turned to straightforward, Soviet-inspired repression.[5] Not only did the defenseless kulaks come into the line of fire, entire villages did so too. But as the peasants failed to oblige and turn against each other according to the class boundaries marked out by Lenin, the Communist Party put the blame on incompetent strategists and members of the Iron Guard movement who had infiltrated the countryside: "Apart from the militia, there was the solidarity of the working classes. There were cases when a kulak family would be evicted and left on the roadside, and the laborer peasants would show compassion and would side with them. There were such cases in Odorhei county, where the entire commune joined in the defense of the kulaks. Peasants in 100 horse-drawn carts paraded in the streets of the county seat to demand justice. There was quite an atmosphere;

the poorer peasants did not agree with such methods. [...] With such impolitic measures, one is likely to encourage the reactionaries."[6]

So the "divide-and-conquer" policy failed to create different classes of in-fighting peasants. The project was far too ambitious and ignored one essential fact: The threat of collectivization was the same for all landowning peasants and it was, therefore, not surprising that they should unite against those who tried to take their livestock and their plots. The partisans were able to resist for the same reason: The peasants in the mountainous regions would regard the communist authorities and their merciless quotas as their common enemy. Conversely, in periods when there was no requisitioning, either of land or produce, the peasants learned quickly to denounce their personal enemies as "kulaks".[7] An example was the aforementioned neighbor of Popa from Scorniceşti, who, having a dispute over land with Popa's mother, had no qualms denouncing him for having an alleged kulak father-in-law. Had the land already been confiscated, the neighbor would not have bothered to denounce him. Attempts to engineer social conflict can succeed only as long as insufficient resources can be used as rewards offered to some at the expense of others. Such resources can include land, wherever it is still privately owned, or food in times of major shortages and starvation, such as in the period immediately following World War I in Romania. Housing, too, can be used in this carrot-and-stick game: In the 1950s and 1960s in Romania, for instance, bombardments and the exodus into towns created a severe shortage of housing. Likewise, state jobs can be coveted by those who regard them as secure employment with a minimum of responsibility. Using such resources as incentives results in social conflict on a large scale. But rather than creating neat, clear-cut social categories in opposition to each other, this kind of conflict produces randomly a society in which everybody is virtually against everybody else. This is the classic definition of a peasant society founded on competition for scarce resources. By extending this model, the communist state was able to replicate it on a gigantic scale, creating an entire society of scant resources.

Communism was an ordinary tyranny, but founded on an extraordinarily ambitious ideological design. The subjection of the peasantry by this modern tyranny was the most comprehensive which had ever existed in history. The consequence of the ensuing class warfare was the dissolution of all class boundaries and the emergence of a dog-eat-dog society where ruthless individualism far outweighed attempts at social solidarity. Nucşoara was, sadly, a case in point.

Veterans vs. bandits

In Nucşoara, the total number of families sentenced to hard labor or executed for political offences prior to the political amnesty of 1964 amounted to some twenty-eight, who between them owned around ninety hectares of land. Their holdings were also confiscated when they were sentenced as "bandits". The region was not collectivized, and the communists struggled to develop the state farm at Slatina. As the confiscated lands were scattered plots on top of the hills, a swap was proposed. The state farm requisitioned the adjacent individual plots—on the lands formerly belonging to the family of boyar Paul which had been divided among poorer peasants in the 1945 land reforms; those peasants were compensated with the plots seized from the political prisoners. However, after the collapse of the communist regime in 1989 a law dedicated to the victims of communism cancelled their sentences and provided the return of their lands. One post-1989 mayor of Scorniceşti recalls the difficulties of retrieving such lands, which had changed hands many times in thirty-five years of fragmentation due to inheritance and sale: "It was difficult when they came back to claim their lands. From '58–'59 to '91 there had been perhaps seven changes of owner. It was difficult, because there were over 200 families owning ninety hectares". (Ion Chelu, a former mayor)

The repression had struck randomly. Aside from those who had taken up arms in the mountains, and their support network, dozens of other people received minor sentences simply for having run across the partisans or having offered them cigarettes or a piece of cheese. The reallocation of the partisans' lands was also conducted rather randomly: The land was given to those whose former plots had been adjacent to the state farms and which had been merged to extend the farm holdings. There were the usual exceptions, favoring members of the reallocation commission—the local predatory elite. But many of the final recipients of the politically confiscated land had not been involved in any conflict. Some saw their own relatives imprisoned:

Names such as Berevoianu, Adămoiu, and Berja could be found in both camps. Initially, the new owners had nothing in common apart from their proximity to the state farm, but later they became a more cohesive group, united by their shared interest. In the thirty-five years after expropriation, the lands which remained under private ownership were fertile and productive. In contrast, the lands taken over by the state farm became, as the peasants explained, "waste-lands" as a result of mismanaged "exploitation". Those who had been the unwitting beneficiaries of the expropriated lands realized that they had to join forces when the political prisoners were amnestied in 1964. None of them did anything to help these people, who had lost everything, although their situation was desperate. Some had lost their homes, as well as their building plots. Vică Berevoianu was put up by his former wife. Constantin Paşol was advised by the Securitate to rent a room from the current owner of his own former home. But the real difficulties started twenty-five years later, when the land swap had to be legally effected in reverse. With the exception of the priest and Petre Ungureanu, everybody else was reluctant to return land. They preferred to go to court, where lawsuits lingered for years. And they all lost, although they had the support of the local authorities and of all those who had been, in one way or another, the accomplices of the communist regime. During the trials, they found a common cause as a group. They called themselves the "Veterans", because among them there were a few World War II veterans, who had received land from the property of the Paul family in 1945. They sent memoranda to the courts, in which the story of the village was told from their point of view:

MEMORANDUM TO THE PRESIDENT

MR PRESIDENT

BEEING A GROUP OF VETERANS FROM THE COMMUNE OF NUCŞOARA, ARGEŞ COUNTY, WE ARE WRITING TO DRAW YOUR ATTENTION TO THE FOLLOWING: As a result of the two world wars, in which several citizens of this commune participated, they were allocated lands within the perimeter of the commune, lands which they owned only for a short time, because the newly created IAS Slatina state farm took them away to extend its holdings.

In the meantime *a small group of citizens allied themselves to the Arsenescu-Arnăuţoiu Iron Guard gang which had been sentenced for the offences they committed.*

These citizens had their lands confiscated following a court order. These plots became the property of IAS Slatina, but because they were spaced out, the farm had no use for them. Therefore, to replace the lands taken from us, the war veterans, by the farm, we were given instead these lands confiscated from the detainees, which we have now owned for some thirty years. On these lands, people have built homes, cattle sheds, and orchards on the produce of which we now depend.

Following the recent land regulations, these detainees are now insistently claiming these plots which we have owned until now; taking advantage of Law 118 of 20 March 1990, they have forged claims dossiers and now wish to take over land legally confiscated and owned by IAS Slatina in Argeş county.

It would be highly unfair to us to have to abandon these lands which we have cared for, and either become shareholders or owners of small plots of poor-quality, weed-infested land. It is not acceptable that these well-cared-for lands should be given to detainees who have made no contribution to their upkeep, lands which we the veterans have been looking after and which otherwise would have been as run-down as the lands of the farm. They probably like the way these confiscated lands look today, but we, who have fought for this country, are not going to let these lands go, whatever the risk.

We have sacrificed ourselves for this country and some of our brethren fell on the battlefield; others came back home wounded and maimed for the rest of their lives. And *this sacrifice cannot be judged in the same way* as the offences committed by the Arsenescu-Arnăuţoiu gang which lured some of our fellow-villagers away.

Their actions materialized into stealing food and livestock from people in the village to feed the gang. Moreover, they killed villagers and soldiers of our country who had nothing to do with the Communist Party; they were simply serving in the army.

The real heroes are these veterans who defended their country and not those detainees who ganged up and fought for power. We request that the situation be examined carefully and that justice be done.

In return for state farmlands, some of the veterans did receive plots which had been held in reserve for the former popular councils (the property of the Church, of schools, or other expropriated lands for which claims can no longer be made following the land regulations). *It would be a further injustice for some to remain in possession of the plots, while others are mere shareholders in the farm.*

We therefore write to you, requesting that our situation be examined and that the organs of power take the most appropriate and humane measures.

This memorandum represents the views *of several citizens* from the commune of Nucşoara, in Argeş county. For and on behalf of these, the undersigned:

1. Băncescu Toma... 41. Adămoiu Petrişor

The memorandum was sent to the president and other bodies by Vasile Stoica, the director of the school in Nucşoara and a local councilor, who was himself involved in a conflict over land with Nel Preda, a former political prisoner. The text is worth a closer reading. It proposes a distinction between two opposing camps: the newly formed group of veterans vs. the former political detainees rehabilitated in 1990, although the latter had in fact included World War II veterans as well. The text also compares the "heroism" of the two groups, alleging that, whereas the veterans sacrificed themselves for their homeland, the "bandits" did nothing but steal food and fight for power. In addition, the "bandits" had a "criminal record": the signatories of the memorandum conveniently ignored the fact that the political sentences passed on the partisans had been revoked and chose to treat them as common offenders who had fully deserved their fate. Accusations of affiliation to an "Iron Guard gang" were typical of the Securitate jargon and glossed over the fact that both Arsenescu and Arnăuţoiu had been career officers with links to the bourgeois parties rather than to the Iron Guard movement.

Although they bought into the Communist Party's version of local history, the signatories to the memorandum were in fact highly dismissive of the idea of state property when they compared the neglected land of the state farm with their own well-looked-after plots. However, they warned of the unfairness of a situation where some would have retained ownership of lands expropriated from the Church and the schools, whereas others would have been left simply with shares in the state farm. The concluding appeal for the perpetuation of an unjust distribution of land was made in the name of "humanity" and asked for "the most humane measures" to be taken. The language used by Stoica, an intellectual educated in communist schools, was the wooden tongue taught to everybody in political indoctrination classes. But though he was the only writer of the memorandum, the feelings expressed there were shared by the entire group of peasants on whose behalf he wrote. Some of them were agents of communization in the countryside, and we shall return to them later. Others, however, were not. Two hundred families involved in land redistribution in a commune of only 1,800 inhabitants is a lot. This would explain why the commemorative cross for the partisans could never be placed in the local cemetery. The fact that people came out en masse to prevent the placing of the cross shows that they all shared the spirit of the "Memorandum to the President". With the exception of the former political detainees, nobody agreed that those who had been executed deserved to be commemorated in the cemetery, and the cross was erected in a forest clearing, an isolated spot to which nobody objected. "The village suffered too much because of them, that's why people were against it," forester Ion explains.

The truth is that many today have nothing good to say about the former political detainees, which was not the case when the partisans were active. There is now much envy against these men, because a special law granted them higher pensions and they enjoy attention and social recognition beyond the village boundaries. This reversal of status has annoyed a lot of people. A law passed in 1990 recognized the category of "victims of communism", which resulted in a new, positive public image of the former detainees who for thirty years had been social pariahs. By 2001, when these interviews were being conducted, the former political prisoners had higher pensions than the village average and had reclaimed their former lands. The fate of Elisabeta Rizea sums up this newly acquired popularity. As the partisans' last connection to the outside world, she was arrested and severely and repeatedly tortured: She was once suspended by her own hair, and her scalp came off her skull. Yet, she never revealed the partisans' hideouts. She became the recipient of several high-profile visits, from Romania's King Michael, after his return to the country to President Emil Constantinescu, and was the subject of several popular TV programs and books. However, in Nucşoara itself, many hated and envied her. In the early 1990s, when she tried to explain to her fellow-villagers that it was no longer necessary to vote for a single party, there were people who did not hesitate to jostle her. This fragment of a conversation between the interviewer (Q) and four notables in today's Nucşoara (Respondents 1-4) is revealing:

Q: *Was communism good or bad?*

Resp. 1: I would like to start by saying that under communism many people could afford to go to school; nobody will deny this. For if in the more distant past only the sons of schoolmasters and priests went to school, later there was an equalization. This had both positive and negative consequences. For example, we the latecomers, we were somewhat marginalized by those better off than us; I wouldn't say things were better then, although it was in that period that I had the house built and bought the car, but I worked hard. I don't know whether others with less education worked as hard as me. Whereas in the past everybody would stand up when the schoolmaster went by, today it's only the elderly who greet me; as for the rest, they're all the same. The man who tills the field might say that I made money by cheating others, so *why should he respect me.*

Resp. 2: I more or less believed in communism, because I thought we were all equal and there was some justice.

Q: *Are people proud of Mrs Rizea, or not really? Because I heard she'd been beaten up.*

Resp. 1: Do you think I don't have enemies? We all have enemies, but not in the same way. If one doesn't have enemies, one is not really alive. But if she had five enemies and they persecuted her, that's another matter.

Resp. 3: I don't think they knew what they were doing.

Resp. 1: We have been unable to capitalize on her image as hero. Had we known better, we would have achieved more, but at least we had the main road asphalted.

Resp. 2: For me, people are all the same, and I treat them all in the same way when they come to the dispensary. When I see her on TV, I don't think very highly of her; she doesn't seem genuine to me. I personally was placed in a rather delicate position. Sometime after 1990 I got a phone call. It was my day off, and I was asked to come to the practice because we were due to be visited by the director of the Health Directorate and by the director of the hospital in Curtea de Argeş. I thought there was something wrong; I didn't know what was up. Two ambulances came and they asked for Elisabeta Rizea; they'd come for her. I saw her, she was walking on her own two feet, but when she saw us she fell to her knees and started crying, that she'd been in prison and couldn't walk. Those gentlemen came with precise instructions that she had to be hospitalized and diagnosed. She hadn't told me, as her local doctor, anything about all this. This was why I felt so humiliated. She arrived at the hospital in a wheelchair and with a whole panel of doctors around her and she was placed directly in the care of the hospital's director. Subsequently she checked herself out; she ran away from the hospital. I told her husband—who was ninety-three when he died—not to worry because "Nana Tuța" is in good hands. And he says, "Lady, she's got shit instead of brain; all her life she never knew what she was doing, and still doesn't." Yes, it's true, she had a difficult life, the way they were all treated. I even thought that maybe she found that life convenient, because she could take some advantage from it. All I can tell you is that she was doing quite alright even under Ceaușescu.

Resp. 1: In Ceaușescu's time, there were communist rallies and we used to go to Curtea de Argeş for those.

Resp. 2: Aside from Rizea, there were other people in the village who collaborated with the resistance, people who all served hefty sentences in prison. There were two sisters who hide for five years in the attic of some house. And I asked Marina Chircă why doesn't she feature on TV, and she said "I didn't want the whole country to know what silly things we did". I never understood why. But I did wonder how Marinica got a pension.

Resp. 3: The majority of people were not involved in politics. The peasant way is to mind one's business.

Resp. 2: You asked me what I think about communism, but I want to ask you what you mean by "capitalism". Well, the way I see it, for this is how I was brought up, capitalism was bad, and I mean it, but I thought that if there are a few rich people in a community, then the rest of us will get our share, that's what I thought. For me, life after 1989 was a disappointment.

Q: How do people look upon the former political detainees?

Resp. 1: I consider them to be victims, because some were sent to prison simply for a smoke.
Resp. 4: Some of those who were in prison do not think very highly of Elisabeta.

Resp. 3: There's no consensus among them either; I'm wondering why *Arnăuțoiu's daughter does not speak with Elisabeta.*

Resp. 4: Some are indifferent, others say "Look at her now, she's a TV star".

Resp. 2: I personally wonder why the *war veterans should have smaller pensions than the former political detainees*. They have very high pensions.

Resp. 1: Nucşoara got public recognition owing to Elisabeta, but people won't admit that.

Resp. 2: If it hadn't been for her, the King wouldn't have visited the village, that's for sure.

Resp. 1: Elisabeta is not even from Nucşoara, she's from Domneşti. But she's gifted for PR, other *political detainees who served sentences as long as hers never appeared on TV and nobody knows about them.*

Resp. 1: I'm persuaded that, had Ceauşescu visited, she would've come out on the porch shouting "long live Ceauşescu".

Q: Are people afraid of capitalism?

Resp. 3: Well, people have this fear of going back to *man's exploitation by man.*

Resp. 2: This was it.

Resp. 3: Afraid we might be tied to someone else.

Q: What are you afraid of?

Resp. 2: Of course I'm scared that Mr Paul comes back and takes my land, or Iorgulescu comes back and takes my land, and I'm left with half an acre and not enough to live on.

Q: Did you personally receive land from their plots? You're a doctor on a state salary.
Resp. 2: I'm speaking on behalf of the villagers. People still have land that was theirs. *I personally wouldn't like to have to work on the land of someone else.*

Resp. 4: The boyars sold the land and the peasants had to work in order to be able to buy it.

Resp. 2: Because they exploit me. What else can I do? I haven't got money, so I have to work. Since they were given their plots back, they should have worked them, but some sold the land instead.

Resp. 3: I think it was fair for these people who'd lost their lands to get compensation, but these lands should have been left to those who held them, because they were not to blame for these land swaps. Now, I had some land there and there comes this law which says they're going to take it and give me somewhere else in exchange, so I suffer. In the entire commune there are *over three hundred of us that had to suffer from these exchanges for about twenty* former detainees.

Q: Back in the fifties, were there people who refused to take on land confiscated from others?

Resp. 3: I don't think so. *Look today, when lands were returned, why aren't the* detainees *working them, instead of selling?*

Resp. 4: *When they claimed their rights, that should've been on condition that they were going to work the land, not sell it.*

Q: Well, why, when you buy a car you have a right to sell it on, isn't it the same with land?

Resp. 2: It's not the same. Why should I give away the land that I was toiling whether in the collective or not? It was their land, fine, but why can't they work, why are they selling it? People work for years in order to be able to buy it off them.

Resp. 4: It's a mistake to give so many hectares to whoever is still alive from so-and-so's family. They should get only one hectare, and some land should also go to those that are known to work it. But *they speculated it*, against me, and against others.

Resp. 3: I'm just wondering what will happen to the dozens of hectares that are about to be returned now. Let's say you live in Bucharest, what are you going to do with fifty hectares of land in deep countryside.

All the participants in this discussion had formerly benefited from expropriated land reallocated to them. They believed that the swap should not have been reversed, or that the former detainees should have received just one hectare or just as much as they could reasonably work. As for those who decided to live in town, they should not have been entitled to any land restitution at all. Envy and enmity are directed against everybody. Some say that other former detainees, and especially Arnăuțoiu's daughter, look down upon Elisabeta, which is false. They also affirm that Elisabeta used to volunteer to take part in Ceaușescu's orchestrated rallies, whereas the political detainees were strictly supervised by the Securitate, especially during official visits, and were not allowed to travel, etc. Anything seems to be said against anybody who, one way or another, stands out and gets recognition. The lady doctor is angry because the ambulance came directly from the district hospital for the high-profile patient, and because decisions about Elisabeta's treatment were taken without her being consulted. In addition, she claims that Elisabeta's late husband, who had served a prison sentence at the same time as Elisabeta, made coarse remarks about his wife.

The mechanism of the conflict becomes intelligible in the light of the above conversation. It starts by building up evidence for the "limited good" and continues with an argument about the unfairness of its distribution. Better-quality land is only one such scant resource. Appearing as a star on TV, earning a good reputation, and receiving royal visits are all rare opportunities. In his overview of the features of traditional peasant societies in the anthropological literature, Foster listed among other characteristics a reluctance to give credit.[8] Peasants believe that the individuals who get to enjoy such rare opportunities can do so only by depriving others of advantages drawn from a limited common pool of goods. Moreover, according to this view, it is self-evident that it was not lack of courage or other moral qualities which led to some individuals being deprived of such goods, but the simple fact that others had privileged access. But because such sentiments cannot be acknowledged either in public or even simply before one's own conscience, certain defense mechanisms are required to justify and legitimize them. One such mechanism is context replacement, whereby negative feelings towards certain people are recontextualized in a way which makes them more acceptable and honorable. A second mechanism is the transfer of one's own feelings and their attribution to others. The schoolmaster, for instance, compares himself to others, claiming that they envy him and that, although he has always worked harder than they have, he remains worse off and marginalized. Therefore, he does precisely what he accuses others of doing, i.e. he attributes greater success to lesser merit. In one and the same sentence, he claims that his life was precarious, yet in the same period (when he was actually brought in by the communists to replace schoolmasters who had been arrested) he bought himself a house and a car, which was quite a feat under the communist regime. The woman doctor claims that Rizea derived advantages from her relationship to the partisans, although it is known that she was sentenced to prison and that all her assets were confiscated. She also claims that Rizea was duplicitous, that she was prominent at the pro-Ceauşescu rallies, and concludes by speculating that Rizea was the one who clamored loudest in favor of the communist dictator.

The substituted context in this local tale of opportunistic "haves" and wronged "have-nots" is the conflict between the rich and the poor from the egalitarian propaganda supposed to justify it. It is alleged that in the past the detainees possessed better lands because they had more money and that today they are still better off because the state has given them higher pensions. Communism found a temporary solution for this inequality, whereas capitalism—i.e. the post-1990 regime—reinstated it. This discourse which justifies envy and injustice draws on several basic principles which can be summed up as follows. Firstly, justice is not objective, but subjective, and represents the point of view of the majority. There are only thirty detainees as against 200 individuals who were allocated land; therefore the former are not entitled to the land, even though morally and legally it is theirs. Secondly, if, however, they have to be given lands after thirty-five years, let them have only one symbolic hectare, which they should be legally bound to work with their own hands and be prevented from selling. The sale of the land is denounced, in the spirit of the communist propaganda of the post-1945 period, as profiteering. Land is not merchandise, it does not belong to one owner but, to use the words of the first communist constitution of 1948, to "those who labor on it". Yet, when the new landowning notables are asked whether they till their lands with their own hands, they say that they use "work for hire, because we don't have the time". Thirdly, it is preferable to be exploited by the state than by someone better off than yourself, which is degrading. To work on another man's land is seen as unacceptable. However, hiring hands to work for you is perfectly acceptable. The important thing is to be numbered among those who can afford to hire labor or, alternatively, to have a situation where this privileged category does not exist at all. In national polls carried out in 2000-2002 more than two-thirds of respondents believed that in "today's Romania there is an ongoing conflict between employers and employees". Fourthly, no differences can be accepted if they favor someone else, even when different circumstances would warrant it. The former detainees must have the same pensions as the war veterans and as everybody else. The mere possibility that lands can be settled on some while others would get only shares in the decrepit state farm is in itself sufficient to divide the embattled veterans into two warring camps. This self-serving, self-justifying discourse is far from benign. Mayor Chelu explains that, even when some of the detainee-claimants won their lawsuits, it took six years to enforce the court's decision, simply because the defeated incumbents refused to leave. The regime change of 1996, which brought the first anti-communist regime to power, did not come a minute too soon, he says. The new Peasant Party prefect authorized the "veterans" to take back their former lands from the state farm, which until then had been fiercely defended by the postcommunist authorities; and so the veterans finally accepted that they would have to swap lands with the former detainees.

It was not only villagers who were successful in manufacturing local conflict. Efforts were made by the Securitate itself and by para-Securitate bodies to inculcate conflict in the very midst of families. When Vică Berevoianu was arrested, for instance, his house was confiscated and given to a relative of his. Later Vică had to rent a room from this man until he was able to have a house of his own built. He also had to make it up with his wife, who had divorced him in the meantime. Significantly, informants recruited from within families either by blackmail, violence and terror, or by promises of wealth redistribution were called "deep-cover informants". Upon their return from prison, Titu Jubleanu's daughter and her husband found only half of their house—her half—available. The husband's half had been confiscated. The new residents were encouraged to take them to court over ownership, although, as fellow-villagers, they were aware of the background to the case. When the political detainees returned, nobody in the village was willing to risk helping them. The Securitate secured jobs for them in places where they could be easily monitored. "When we came out of prison," Verona Jubleanu recounts, "we had to go to court for the other half of the house. They wanted to bring a boy from the village into the house. We had no-one to help, there were some relatives who would bring food through the back-door, my sister, my little brother, and an elderly aunt, my father's sister, who had looked after them. Nobody else." In a separate case, Securitate reports described tersely how two men known for their ties with Arsenescu were discovered hiding in bunkers which they had dug under their own houses: "On this occasion, we used only deep-cover informants," the reports explained.[9] In other words, they were given away by their families.

People without much education, such as the Rroma (Gypsy) handymen, cannot deliver elaborate, self-justifying speeches in the manner of the village intellectuals. They worked for the cooperatives in the plains until 1990, minding the cattle in exchange for maize and other produce. When the cooperative farms were dismantled, they were left with nothing to do. As they had no land of their own, they were allocated some under Law 18. Not on the site of the state farm, which even during the transition was regarded as state property, but from the detainees' lands. In 2002, Ioana Arnăuţoiu, Toma's daughter, was still fighting in court to reclaim her land. The fact that these particular lands were chosen to endow the Gypsies is no accident, but derives from the old tactic of pitting one social group against another. The domination mechanism chosen in this case involved turning the conflict between the state and a group of citizens into one opposing two groups of citizens, so that the state could pose as a referee between the two. Engineering social conflict is essentially an economic mechanism of domination whereby the state turns a process which would otherwise necessitate investment in enforcement into a spontaneous process, and mobilizes private energies for the benefit of those who manage the conflict. In the early years, the communists recruited massively among the marginals. In the district committee, says Gh. Şerban, there were enough Rroma and people without an education. Things were no better among Securitate personnel. The former detainee Constantin Paşol has this story to tell: "There was this elderly colonel from Piteşti, he was from artillery, and a Gypsy who worked for the Securitate was beating him up; he would just smash his face every time. The Gypsy's name was Zamfirescu, he was with the Securitate, but before that he had been a servant in the colonel's house. There were three of us in one bed, one with a broken leg from the war, the colonel, and myself, and many others on the floor. For food we had 100 grams of stale polenta, that's what we had every day, that's how we lived for five years. After Piteşti, we went over to Jilava prison, then to the moors of Brăila to cut thatch."

After several hard years in prison, Constantin Paşol returned to the village to find his home confiscated, his land taken, and the place full of "a lot of Securitate people, very much like today". His was the strangest story of all, as suggested by a comprehensive table in which the names of those whose lands were expropriated are listed on the left and the beneficiaries on the right. Next to his two plots—of 0.06 hectares each—listed as belonging to a "Paşol Ctin", one in the village of Topile and the other simply named "home", the same name is written on the right column and on the left. There was a second "Paşol Ctin", one of Gh. Şerban's first recruits, who became Communist Party secretary and who, as a member of the land distribution commission, simply helped himself to land. He had had an old dispute with Iancu Arnăuţoiu, the brothers' father and an influential local man before the war, and joined the communists in the hope this would make his position stronger. After 1990 he fought long and hard for his land, but, of course, he had no rights over it, so he lost. Today he tells his story grumpily, without a trace of the sense of humor he showed when thirty-five years before, as a party apparatchik in the land commission, he decided that the plot he would seize for himself would be that belonging to the other "Constantin Paşol".

Although highly spectacular, Nucşoara's case could be dismissed as exceptional or unrepresentative, but nothing could be further from the truth. In the communist period such redistributions of land were common, although admittedly they were less frequent in the countryside, which had been almost entirely collectivized. The principle enshrined by Law 18/1991, which required transferring lands to those who had never had any from those who had, merely expanded the application of this communist logic. There was no rationale which might have explained why a European state—albeit a peripheral one —would create new landowners in a country where forty per cent of the population had already been employed in agriculture, instead of drastically reducing that number. Likewise, postcommunist Romania's main social conflict had a similar source: It pitted former property owners, whose houses had been nationalized by decree, against the tenants forcefully relocated in their homes by the state. The postcommunist state pretended to be refereeing this conflict, but the Tenants' Association was widely known to support the electoral campaigns of the postcommunist parties. Families in their hundreds of thousands were involved in the conflict. A tiny minority of these was formed by the profiteers who lived in confiscated luxury villas, but they were people with strategic positions. A report dated July 1997 by the Prime Minister's Audit Office[10] showed that the majority of these villas were occupied by politicians. Some, like Ilie Verdeţ, the man responsible for the repression of the miners' revolt in Valea Jiului in 1977, dated back to Ceauşescu's time; others were members of the postcommunist parties. The state's manipulation of the conflict between owners and tenants was duly penalized however: In the first twelve years after the collapse of communism, the Romanian state lost all the lawsuits over property referred to the European Court of Human Rights in Strasbourg. But while there were few people willing to pursue justice as far as Strasbourg, the strategy brought double benefits to those who orchestrated it. Firstly, they benefited from the votes of the almost one million tenants, who, in Nucşoara and elsewhere, far outnumbered the lawful owners. Secondly, members of the predatory elites continued to live in, and gradually started to buy, such confiscated houses. The main beneficiary of invented social conflict has not been the communist state, which, after all, is an abstraction empty of content, but whoever happens to hold the state captive at any one time: the predatory elites.[11]

Notes

[1] Cătănuş and Roske, *Colectivizarea agriculturii*, 69.
[2] Stalin, *Opere*, vol. 2, 106–7.
[3] Fitzpatrick, *Stalin's Peasants,* 29–33.
[4] Seton-Watson, *The Pattern of Communist Revolution*, 45.
[5] Ibid., 157.
[6] Cătănuş and Roske, *Colectivizarea agriculturii*, 166.
[7] Fitzpatrick, *Stalin's Peasants*, 16.
[8] Foster et al., *Peasant Society*, 313.
[9] Oprea, *Banalitatea răului*, 287.
[10] The Valerian Stan Report.
[11] The concept of state capture was first coined by Daniel Kaufman of the World Bank.

CHAPTER 5

The Destruction and Replacement of the Elite

"It must needs be that offences come, but woe to that man by whom the offence comes!" These words of wisdom from the Gospel of Matthew (18:7) are recurrently invoked by the old former political detainees from Nucşoara, who are deeply religious peasants. One can spend hours listening to the catalogue of woes which struck those by whom offence came: cancer, heart attacks, car crashes, and even lightning. One of those who tortured Verona Jubleanu was actually struck down by lightning as he was opening his door. A godson of Iancu Arnăuţoiu who refused to allow his godfather to return to his own home when released from prison was also struck by lightning, under the open sky. Elisabeta Rizea and Nel Preda, too, had their panoply of the devil's disciples struck down by divine will.

In Scorniceşti, too, the former political detainees point proudly to those who pursued the collectivization policy, now all dead and buried. But they are alive, having survived the gulag to reclaim their lands and their tractors after the fall of communism. Outsiders with less faith or even insiders, who lost their faith in prison when, like the former detainee Paşol, they were made to urinate in their own boots for days on end, may be less convinced that justice has been served. To use Paşol's words: "Well, times have changed, but not for the good, you know: They're still after us. Look at this woman from the mayor's office, the secretary; she's a nasty piece of work, her father was in the Securitate, and she's now running the village; and there's the other Securitate fellow, one Ungureanu by name, he's now a councilor, and it's them in all the local bodies." It is hard to disagree with Paşol. As you stand on the terrace of Şerban, the first mayor under communism, who profited from the expropriations of detainees, his land and his family's fill the horizon, and they used to be the poorest in the village. His son—the man formerly in charge of the cooperative delivery quotas—has installed in front of his imposing, brand-new villa an enormous satellite dish. It is quite true, however, that things could have turned out better for Emil Bărbulescu, son of Lina and nephew to Nicolae Ceauşescu. Sulky and unshaven, he brings his troubles to anyone who wants to listen. The electricity company cut him off for unpaid bills; and his ungrateful fellow-villagers forgot that it was his parents, not Ceauşescu, who turned that village into a town. And yet, comrade Neacşu, mayor of Scorniceşti during the systematization and re-elected by the villagers in 1996, when the national presidential elections went to the anti-communists for the first time, appears to be doing quite well. He is in no danger of being hit by lightning as he clumsily sits down in front of his computer and starts up the server—a gift from the European Union—to monitor income tax in his new position as financial director for the mayor of Scorniceşti. It is a job for which he volunteered in the run-up to retirement. But maybe the town is protected from righteous lightning and other such hazards by donations made to the local church by the mayor and his wife, two youngish communist upstarts prior to 1989, successfully converted into capitalist entrepreneurs. They are the latest in a long series of church donors which now includes boyars and Communist Party members alike.

Three sets of historical factors have been identified as favoring the installation of communist regimes: the party's own strength, outside military intervention, or a combination of both.[1] Romania is generally considered to belong to the second category, given that a genuine Communist Party did not exist there at the end of World War II. Aside from cadres imported from Moscow and agents of direct repression, the new regime needed insiders to work insidiously towards its ultimate objective: social change. These agents of change are the topic of the present chapter. Their character changed with the times, as the Ceauşescu era gradually transited towards a second stage in the communist regime's control over society. This was much more sophisticated than the previous Stalinist one, based on mere coercion, which ended by 1964. That year, all the surviving political detainees considered non-dangerous were amnestied. (Ion Mihalache himself had died in prison a few months before the amnesty.) In the second stage, control was exercised by persuasion rather than coercion: The state gave to, as well as took from, people, and it presented both options to the peasants as the two extremes of a continuum. In order to benefit from the many rewards, chief among which were the benefits offered by the welfare state, one had to meet a single requirement: belong to the system. This could mean party, cooperative farm, or trade union, depending on individual circumstances. This arrangement was more subtle and less costly and had the advantage of showing to the outside world that this was a society in which communism had acquired legitimacy. Coercion was now used only in moderate doses. Ceauşescu's Securitate used surveillance and bullying, but penalized less than the political police of the Stalin era. There was no need for punishment any more. Some 1,500 individuals were condemned for political crimes in Ceauşescu's time compared with 300,000 in the Dej era, a ratio of 1:200.[2] Society was communized successfully, and that still showed in the first few years after the official fall of communism in 1989.

Traditional authority

Hugh Seton-Watson[3] has shown that, in underdeveloped countries the Communist Party recruited its cadres especially from poorer peasants and intellectuals, and in advanced countries from workers and to a lesser extent from the intelligentsia. Once selected, these people became "classless" as they advanced into a special, privileged apparatus category. The sole support of repressive forces is not enough for the agents of change to be able to assume power. Power based on coercion alone can dispense with legitimacy only to the extent that all other sources of legitimate authority are destroyed. Even once its political influence has waned, alternative authority is intolerable to coercive regimes. But in Nucşoara—more plural than Scorniceşti by virtue of the greater autonomy of peasants in mountainous areas—this alternative authority existed. The area had its own social entrepreneurs, its local brokers, who were all the more influential as their authority was recognized by the peasants. These were people like Şuţu in Domneşti, Elisabeta's uncle, the owner of a sawmill and shops that employed many in the community. He also christened the children of his social inferiors and was a model in everything, including electoral choices. He was so influential and highly respected that his arrest would have been highly problematic. They could find no reason to arrest him, so they had to shoot him, explains the second Paşol, the communist.

Iancu Arnăuţoiu, father of Toma, was less wealthy than Şuţu, but nevertheless influential. The son of a priest, this local schoolmaster, well-remembered by his pupils ("The history I still know I learned from him," says Eugen Popescu), had over time become a small landowner. He brokered many business transactions, he was a real local entrepreneur. Costică Paşol's assessment is accurate, though full of resentment. "He was a politician, too. He used to have a say in the nomination of mayors, and usually had his own men elected. He had a lot of godsons locally, who, in spite of their poverty, would ply him with gifts in exchange for favor," he says. Another heavy-weight entrepreneur, much sought after for his advice and support, was Nicolae Niţă, Toma's father-in-law. The fact that so many came to seek his advice led him to believe that he could put together a group of partisans and give its leadership to his son-in-law, a career officer discharged from the army.

Before the peasant reforms of Prince Cuza, Nucşoara had been one of the estates which belonged to the monastery at Curtea de Argeş. This is why there were no boyars in the area. Paul and Iorgulescu, the holders of the oldest boyar titles in the region, had bought their land, not inherited it, at the same time as many peasants became owners. The old rural landowning families did not profit from Cuza's secularization of monastery lands: Entrepreneurs with capital did instead. Even if they were wealthier than others, the local landowning elite members did not have absolute control over resources. They brokered business transactions, guaranteed credit, made loans, offered protection, arranged marriages and christenings, and generally posed as patrons, but they had no monopoly of any one resource. They also belonged to different political parties: The region had its peasants, liberals, and Iron Guardists. The situation was the same at Scorniceşti. Leadership alternated. That the market economy worked is illustrated by the gradual downsizing of the oldest "boyar," Iorgulescu, who was eventually forced to sell land. The peasants could afford to buy: Fifteen families bought the Pojarna mountain from Iorgulescu. The communist Costică Paşol himself was able to purchase land "before the war, with a loan from the boyar. I thought I might be able to live off a small bit of forest, so I bought five acres." Now, five acres is not exactly a "small bit." Therefore, in no way could the local elites, priests, schoolteachers, and petty boyars be regarded as gatekeepers, i.e. as elites controlling access to common resources, even though they might have played a considerable role in facilitating access. They were merely social brokers.

In the semi-literate rural world of 1945, the priests and schoolteachers had another, and possibly more important, role to play. As the only educated, world-savvy men in that milieu, they were the custodians of public morality, the molders of village opinion. They signaled what was right and what was wrong. If words uttered from the pulpit and in the classroom coincided and strengthened each other, it became difficult to give any credence to the party activist, especially when, like Şerban or Paşol, he was a nobody with barely five years' primary schooling. The activists had little land and even less credibility: they had spent time in Russian prisoners' camps or in Russian-created army divisions and were, therefore, compromised as Russia's agents.

The communist regime had two alternatives in dealing with the village elites. One would have been to win them over, but, as they were patrons of factories and owners of land, that was not an option. The communists bought only people who possessed nothing, usually intellectuals. The second solution would have been to have these elites destroyed, given the fact that they were the support base of real authority, of the traditional parties and the monarchy, the pillars of old-regime Romania. Later, it was this maximal solution which was adopted. In the case of the Orthodox Church, it was applied very thoroughly: In a historic compromise, the Church allowed itself to be plundered of the remains of its former wealth and to be made dependent on the state.

Nowadays still, good reputation and prestige—difficult-to-replace capital in the countryside—are a headache for the agents of social change. What Costică Paşol finds particularly hard to swallow is Arnăuţoiu's reputation:

> After all, who was this Arnăuţoiu? He was the greatest crook. What did he do? He was a schoolmaster who beat his pupils to a pulp. We had a bank, "Banca Albina," he bankrupted it and with that money he bought Paul's estate in Slatina, which he then sold on to the villagers at a profit. Iancu Arnăuţoiu acted as proxy in the sale of lands

> in Slatina, Brusturi, Ogradă, Lazului, Dumitean. The peasants always lost out, he cheated them. Up to World War I the peasant used to work as a tenant on the boyar's land; the boyar would give him a small plot, but alongside this he had to work on the boyar's land, too, otherwise he would lose his tenancy. From his mediation of sales Arnăuțoiu was left with loads of land, sixty acres himself. And another thing: People around here used to have sheep, and Iancu Arnăuțoiu had some forty acres of grazing land near Mount Zănoaga, at Belciu, another mountain not far from Zănoaga. He managed things there so that peasants did not produce more than one kilo of cheese all summer.

In spite of such efforts to discredit him, Arnăuțoiu remained popular, supported by his broad patronage network, chiefly of priests and schoolteachers, and was widely respected. People respected Toma as well, because he had education and he was an officer. Even when the Arnăuțoiu sons were hiding in the mountains, people would still ask Arnăuțoiu senior to be godfather to their children. They thought that the regime would change and he would return to power. The wife of the man accused of working for the Securitate, Ungureanu, had been baptized by him, and the husband claims this as the reason why the partisans spared him, although he came within their shooting range a few times.

For many years, as long as there was hope that the "Americans" would come to liberate the country, people oscillated between these traditional sources of authority and those imposed upon them by the new coercive regime. Later, and gradually, it became clear that there was only one source of power, and that far from representing an alternative the partisans in their mountain hideouts were just a group of hunted people. After the failure of the Hungarian Revolution, things became even clearer. The old elites and the people supporting and respecting them had become little more than quarry, and there were considerable rewards for those willing to help hunt them down. Besides being offered rewards, those on the wrong side of the accepted social norms were given the opportunity to redeem themselves. The Arnăuțoiu brothers were sold out by a friend of theirs and a member of the local gentry, whom the Securitate were harassing because he was homosexual.

The descendants of the old families saw their fortunes decay, and they are no longer living in the villages. In Nucşoara, the repression destroyed two generations of the same family in some cases. The descendants of Coculescu from Constantinesti are now living in Paris, those of the Lînaru family from Scorniceşti in Germany. Gigi Paul, the son of the former landowner from Nucşoara, is now living in Bucharest and has had to struggle to reclaim a manor that his family had owned in Argeş county, at Calinesti. Neither have reclaimed family land, but the mayor's office in Scorniceşti was worried about an application for a house by mail from Germany. Of Iancu Arnăuţoiu's grandchildren, Toma's daughter is now living in Bucharest, while a cousin of hers has returned to the village after winning back his land. However, he keeps himself to himself. Şuţu's grandson is the only one who entered politics, and, like Ion Berja, the son of the political detainee, he ended up leaving the Peasant Party for a postcommunist party, though, unlike him, he never became mayor. He tried his hand at business too, failed, and is now living like a peasant, proud of his plum brandy home distillery. But at least he went to university. Others, such as the sons of the priest Drăgoi, were barred from university because of the political standing of their families. The party made sure that no-one in the second generation could stand out, not even as mere intellectuals.

After 1989, the villages which composed Scorniceşti tried to escape the false identity Ceauşescu had provided them with, and found some sources for a genuine identity among the old-regime families. The association in Mogoşeşti called itself the "Dr Voiculescu" Association, and its ill-fated counterpart in Constantineşti took the name of the Coculescu family. However, such legitimizing strategies have nothing to do with law or justice. The prospect of any of the former owners returning is not being encouraged, especially by the agronomists who run the new farming associations, having taken up the old boyar manors and even using their name. Communism had stripped the old elites of everything but their names; transition achieved that.

The hunters and the hunted

The agents of change—i.e., those destined to become the new regime's elite—were recruited in several stages. It is worth remembering that the communist regime lacked party cadres in Romania: The mass opposition to the political establishment was provided by the Iron Guard, an Orthodox, populist movement which claimed one million members and was close to the peasantry. Therefore, in the initial stages the communists recruited former prisoners of war from the Eastern front, especially the more vulnerable ones, without much land or education, such as Şerban and Paşol. Paşol recalls the circumstances: "In '41, I was on the front, we'd crossed the river Prut, and we fought in Russia for quite a long time. In '45 I came back from the war. I wasn't into politics, I had no idea what that was. No, sir, to be involved in politics, one needs some education, one must know at least what politics is. After the war, Şerban came along: he'd been a prisoner in Russia, in Siberia, and he returned with the 'Tudor Vladimirescu' troops. He became mayor here, and one day he comes and says: 'Costică, you must join the Communists.' 'Alright,' I said, 'I'll join.' So in '46 I joined the Communist Party."

Gheorghe Şerban too recalls how he, a poor peasant with hardly any schooling, became mayor of Nucşoara:

> Arnăuţoiu and Drăgoi were both schoolteachers, Arnăuţoiu was my schoolmaster. Father had a small house built, 'cos he didn't have a lot of land. The land belonged to boyar Iorgulescu, who owned seven mountains and was deputy for Câmpulung.
>
> They [the Russians, *author's note*] took me prisoner in Odessa, and I stayed there, I don't know for how long. A long time. I was detained in seventeen camps, all over Russia. They'd get you and take you someplace else; I was in Vladivostok, in the Pacific, we used to stack the fish into barrels there. I stayed there for a whole year, then they took me to Siberia. There were 3,000 of us prisoners, and we were kept in an underground camp with guards in the four corners, and the barbed wire was four meters high.
>
> Some time around 1945 I returned to Romania. I arrived in the autumn of 1945 and someone told me to go and join the regiment I was in

when I first went to war, the Thirtieth Infantry Regiment in Câmpulung. Then I returned home. A few months later, someone from Câmpulung drove up in a car and called me to the mayor's office. The accountant was there, too, the cashier, and I didn't stay long. He said to me: "As of this moment you are the commune's mayor. You'll get the nomination from Câmpulung in a few days' time." I didn't know the man and I still don't know who he was. He was with the Communist Party. How could I know anything when I only did five classes in primary school? The man at the mayor's office, the accountant, he was a neighbor and he helped me, he made the phone calls on my behalf, made payments, he helped with everything. How was I to know about state administration? I was just a simple man. It took a while before I learned how to use the telephone. [...] Then they made me the head of the local administration in Domneşti for about a year and I had sixteen districts under my management. But a communist? How could I ever be communist? I didn't even know what the word meant. I didn't have that kind of understanding.

Gheorghe Şerban had a long career as a communist civil servant for Muscel county, at district level and in the party apparatus. He is full of contempt for the assortment of people brought to serve in the local administration:

They had been mopped up off the streets. One of them, a cobbler, he was made the chairman. And there was another one, a Gypsy, for this is what he was [...] he was a Gypsy [...] That one became deputy chairman. That provisional committee was formed of five people: chairman, deputy chairman, and three members. And so they installed us there, they had a meeting in Câmpulung, and they elected us to the provisional committee of the Muscel county prefecture. There were five of us and we were given a car, and a driver. We would go round the mayors' offices in the county. This happened before they started forming the farming associations. They appointed me in Domneşti and gave me a horse-drawn gig and a secretary, and then they appointed me prefect in Câmpulung, where I remained for four years. They sent me to a school for political officers. I did one month's training and then I was sent back to Câmpulung. After that, I became an employee in Topoloveni, where they started the first collective farm. It wasn't easy because there were a lot of kulaks there. I went round trying to get people to join the collective. I've got thirty-one years' experience working in state administration.

I stayed in Câmpulung at the prefecture for three years and in that time a lot of collective farms were set up. Then they sent me to Topoloveni where they made me head of social services. After that, they sent me to Curtea de Argeş, where I was head of the Section for Communal Farms and Local Industry for Argeş county. Then I was promoted to head of the Trade and Business Division; I was in charge of commerce throughout the county. I had a car, and the Communist Party secretary used to call us round to make presentations and give reports. I was in that post for a few years and they called me back to be mayor in the commune for a second time. I remained mayor for several years, I don't know, thirteen-fourteen; it's all recorded on my employment card.

[...] It was the fault of the prewar parties. Father Drăgoi and the schoolteachers got together and got organized. They [the partisans, *author's note*] got organized in the mountains; they would go hunting and would take sheep from the shepherds, but would give receipts for them because they planned to pay them back once they came to power. They shot an informant who turned in the mother of one of the partisans. The woman went one evening to a relative to ask for food. And the relative, Băncescu by name, had her sleep in his house, and as she lay sleeping he went to report her to the militia: "Listen, I've got the mother of one of the folks in the forest with me; I gave her food and a place to sleep." The militia came with reinforcements and caught her. Then they [the partisans, *author's note*] shot the man, but he survived, he didn't die.

[...] The Securitate tactics were not great, but in the end they caught up with them. They brought in the army and they were all over the mountains. When they saw so many military in the mountains, they went over to Corbi and built a sleeping shed in the ravine at Poinărei. From there they had a view over the main road, to the bridge over the river Doamnei, that's why they chose the place. Almost four years they stayed there, but the army and the people smelled a rat and found them out in the ravine. And they caught them in the forest and the Communist Party's Securitate got them all. Two were caught in the middle of the village, then the rest too were all caught by the Securitate and were taken away. They didn't shoot them, they needed information on the others.

[...] I didn't get involved. I was mayor twice. I was a simple man but I had brains. No, the Securitate would arrive and the party activist would tell them where they all were, go there, and there [...]

When they got the folks in the mountains, the Securitate came to me and asked for the farm labor register. They asked where the register was and they took the labor register over to Piteşti and kept it there until they identified all of them. There was one who hadn't been with the others in the mountains, but he had collaborated with them and they confiscated his wealth. And then, once they identified them all, they only took the men's land, not the women's.

[...] When they tried these folk from the forest they didn't call for me. I asked a Securitate man why didn't they call me as a witness at the trial, and he said that I hadn't had anything to do with them, but he also said that they said at the trial that they could've shot me when they ran into me.

The village marginals thus recruited received a smattering of education. Costică Paşol did a month's worth of party training and he still regrets today that he did not go to university. But he made good use of his training and spends his retirement reading Marx's *Selected Works*—"a great book, Marx's father was a big industrialist, you know, and he spoke about 'surplus value' there."

That was the first generation. Later, the party started educating its cadres, as they had a desperate need to replace the old-regime political enemies. The first generation of new village intellectuals was rushed through schools, and many were admitted to higher education on the basis of their "sound" social background. Most often, they would be people who already worked and who had completed "open studies" degrees to fill posts in the local councils and schools left vacant after the elimination of the old-regime civil servants and intellectuals. Many saw in this an opportunity for rapid upward mobility and took full advantage of it. The present elite are the direct descendants of that generation.

Resp. 2: You see, my father was a party member, I was a member, and I still stand by it. I think my father, too, joined the party for our sakes, to make our life better.

Q: And what did your father do in the party?

Resp. 2: Not much, he just went along to meetings. He worked as an accountant at the Retail Cooperative in Brăduleţ. He started as a

> cleaner at the age of fifteen, then was an employee at the mayor's office, then secretary. He went on to do "open studies" college and then he became an accountant. I studied medicine in Timişoara." (Focus group graduates, Nucşoara)

On the list of confiscated plots, the right-hand column under "recipients" includes almost the entire new village elite. In the left-hand column are those from whom land was taken, all the old-regime priests and schoolteachers. The mayor's son took land from Father Drăgoi, his neighbor, and it was on that plot that he built his villa and installed a satellite dish. Many had houses built on the confiscated plots, in order to mark their territories. The wife of Şerban Jr. is a secretary at the mayor's office and has held this post, with minor interruptions, for longer than twenty-five years. She is the one who keeps under key the Land Register, which contains all the title deeds. Other village intellectuals—who, like her, took part in my group interview on the subjects of communism and of Elisabeta Rizea—are all members of the local council, Nucşoara's mini-parliament. They, naturally, represent the postcommunist Social Democratic Party, which won all the council seats in the village in the local elections. "All the intellectuals in our commune come from neighboring villages, they married some rich girl from the area 'cos they were poor, and then they all filled the good teachers' jobs and the Communist Party. The Securitate and the communists—even today—they form the nucleus of the main Social Democratic Party here." (Nel Preda) In the circumstances, the former detainees had to fight long and hard to get their lands back, even when the law was on their side after 1989. These people, branded as "enemies of the people" under the communist regime, were still considered as such by the militiamen repackaged as policemen in postcommunist times:

> The newspapers said that, on the basis of a 1990 Parliament decision, the former political detainees were going to get their lands back. We didn't sue anybody for our lands, the state did it on our behalf; the final outcome was announced a year and a half later. Vasile Chelu owned my former land and had fenced it, so I went and asked him politely to remove the fence and hand the land over. But before he could remove the fence, this policeman, Dascălu, comes along and I ask him: "What's the problem? I've got a court order."

The next day, his deputy, one Şova by name, came and took me to the prosecutor's office, and placed me under arrest. He took me before Prosecutor Pelinescu. And the prosecutor says: "Well, what are you doing? What's this, vigilante justice? Can't you wait until you're given legal possession?" And I say: "So who's going to give me legal possession? I've got a court order, but the council are dragging their feet over it." I took out the court order, the prosecutor wrote something on a piece of paper which he then put in an envelope and gave to the cop: "You go to the mayor and you give him this; tomorrow he'll go there and grant this man possession of the land." Now he's deputy head at the station in Domneşti, this cop. (Vică Berevoianu)

Nel Preda adds:

These people were planted here. In our commune they were placed in key posts. I served a twenty-five-year sentence and had two houses confiscated; they even took the bricks, and all the land that I had. They confiscated my five stables. And now these communists want to take the land I own by the lake. The mayor of Nucşoara had my storage room demolished; he was a cobbler, like Ceauşescu; Chelu was his name. I had to pay rent when I came back home. I lived in Sboghiţeşti; the mayor there was Şerban, the man who goes around wearing tight peasant-style trousers. After 1990, it was our turn for power, but no, it was still them that got all the offices. All these people are from the commune, and they all made money; one of them who did well out of stealing bits and odds from the cooperative farm was and still is with the communists, and did well out of them. They don't realize that all their thieving and all their scams were at our expense, mine and other citizens'.

[...] They wouldn't give us our land back. I had to do the rounds of tribunals for four years. They referred the case to a higher court at Curtea de Argeş and I won there, then they appealed and the case was heard in Piteşti, and I won there too, and then they gave me the land back.

[...] The secretary, the Şerban woman, she's secretary and farming agent. A few years ago this woman lawyer with full training was hired at the town hall, but she could not straighten things out; the others wouldn't let her do her work properly. The councilors were all forewarned and they sided with the Şerban woman. She gave land to all of them and now they're all supporting her. I had an argument

> with her. That was when my title deed first arrived from the prefect's office. I tell her to place me on the list of land entitlements for I had won the land back legally. And she says: "Listen, comrade, this is a forgery." Now, how could a document signed by the prefect be forged? I served my sentence in the same prison with some important people, and I am not impressed by an impertinent village clerk. This happened a few years ago. This is what happened: my land was given to a school director from the area. He then made a swap with some kinspeople of the secretary's. She told them to stay put on the land as there was nothing I could do against them. I petitioned Parliament saying that Mrs Şerban is stopping me from reclaiming my land. Parliament responded, but the letter was also sent to her at the mayor's office.

The battle is still on, it would appear. One group, the persecuted, has the law on its side. The others, the profiteers, have the state. Vică Berevoianu was arrested by the police, for whom he is still the source of trouble, as he was under the previous regime. The daughter-in-law of Mayor Şerban de facto runs the mayor's office irrespective of the political regime, and any responses to petitions sent to Parliament against her end up at her office. She is also the local farming agent, responsible for the land restitution law. As long as this elite has captured the local state, they will continue to dominate the rest of the village. And it is not difficult to see that the state still identifies with them. In one simple exercise, when people are asked "Who represents the state here in Nucşoara?," the name of Mrs Şerban, the secretary, is the first to come up. The second is the name of Petre Ungureanu, widely considered to be the village chief informant (*securist*), the man who discovered Marinica Chircă in the attic where she had been hiding for five years and delivered her to the Securitate. But Ungureanu is no ordinary *securist*. He arranged for his son to become a policeman, rather than sending him to university, a choice which suggests that he knows exactly where power lies. And being, unlike Mrs Şerban, an enlightened individual, he returned the land he had acquired from a detainee to its rightful owner in an out-of-court settlement. His nephew is a deputy and prefect on behalf of the more enlightened of the two postcommunist parties, the PD.

> How Petre Ungureanu was recruited here in Nucsoara is close to the story of the kulak from Scorniceşti, who ended up being the first to join the collective farm. He was in a group identified as scandal-

> mongers during the elections hijacked by the communists in 1946. Later, as a forester, he shot a stag without proper authorization and was found out. They started pressurizing him and he had to become an informant. Securitate people afterwards became his friends, his allies, his kin. Today, he is widely regarded as the most influential man in the commune. He did not become mayor, but he did not have to, with his nephew having the higher office of prefect. He is content with being a councilor and mayor-maker. For many years, he oversaw the wood and timber resources in the area. Forest land confiscated from boyars and peasants alike became state property and access to it was obtained only through the good offices of the foresters. "Ungureanu had the full use of the woodland, and he chopped wood everywhere," Nel Preda says. "He helped everybody with wood. Poor people worked for him in shifts: some in the morning, some in the evening, and it was "Mr Tică this, Mr Tică that. They're still afraid of him." The old communist Paşol is also fuming: "I know injustice when I see it, and I couldn't stand even our own old communists; they were scoundrels, they stole from the forest and caused all sorts of trouble. They'd steal and sell the wood. Ungureanu had the entire communal forest land stolen, and they couldn't do anything against him, 'cos he was the Securitate's man. I'd denounce him in party meetings, and he tried to have me arrested, but couldn't.

But things tend to be a little more complicated with people who, like Ungureanu, were recruited from the "other camp." His wife's father, suspected of links with the partisans, was beaten up so brutally that he died shortly after returning home. Ungureanu himself was standing next to the cars the night Berevoianu and Preda were both taken away, but he says it was not he who denounced them. In fact, he says, they remained in hiding for many years without anyone denouncing them, but in the end the Securitate and the army infiltrated the place; they were everywhere. He joined the Communist Party upon being discharged from the army, otherwise his children would not have been allowed to go to school. It was in this way that his daughters could train to become teachers. This is how he explains the influence he enjoyed in the commune:

> I've always been an honest man, and while I was in employment, if anyone came to me I'd help them without asking for anything in return, and people remembered me. I did favours to people in a disinterested way: a cartful of wood, an armful of branches in winter, I would charge them less than the normal price. I did have another

> advantage, though, from the communist period. I made friends with that *securist*, Ploscaru; he was only a lieutenant when he first arrived in the village some time in '51–'52. Later he built quite a career; we were close and whenever he came to arrest someone I would intervene on that poor man's behalf, for people had suffered enough.
>
> [...] I talk to people, like this, among family and friends. I sometimes help folk in trouble [...] that's why people listen to me.
>
> [...] I entered the National Salvation Front in 1990, I was chairman in the commune and in '92 when leaders Iliescu and Roman split up I sided with Petre Roman, because a nephew of mine, Cârstoiu, entered politics as a member of the PD. He was a prefect and a PD deputy.

Deep in the mountains where Nucsoara lies, the foresters are the major gatekeepers: They control access to the most important public resource, wood. As one local says: "We've only got two things here, wood and wildlife, and for both you need authorization from the forester." People are authorized to cut a certain amount, but in practice they will cut twice that amount. The mountain-dwellers fear that one day the forest will be gone, yet nobody seems willing to stop the cutting. It all started with the massive forest cutting program which was supposed to help Romania pay its war debt to the Soviet Union in the 1980s. In areas around Nucşoara, the local Forestry Service had an annual forest growth capacity of 20,000 cubic meters, but people would cut around 50,000–60,000 cubic meters, three times the annual growth rate. Large surfaces of forest land started disappearing. Today, the Forestry Service opposes the return of forest land to private owners fearing that people would start cutting wood indiscriminately. Nevertheless, for the time being it is state employees, i.e. the foresters, who have a monopoly of forest cutting. They seem to have a say in social policies too. Chelu, the former mayor, had a word with the chief forester to allow the Gypsy woodcarving artisans to "collect a few bits of wood here and there, 'cos they're starving." During elections, the local social democrat chief, "Mr Cornel," a forester himself, drives the Gypsies to the ballot in his own car, pressurizing them by threatening to rescind their wood-cutting allowance. The villagers asked for the Pojarna mountain to be returned to them, but this has not happened yet. In the meantime, the Forestry Service permits forest-cutting solely in the villagers' section of the forest, which means that by the time they have a chance of getting it back, there will not be much left.

Forest is an essential resource, controlled by the state's representatives. Equally essential are farming subsidies, and all the related paperwork and certificates. For all these, it is very important to have support from the mayor's office. Sică Dumitrescu, who was appointed mayor by the locals after the Revolution of 1989 and worked *pro bono* for a few months before withdrawing, sums up the situation in simple terms: "Everything is run from the mayor's office. Whether it's to do with wood allocations, or the creation of small local businesses, or taxes, anything, one needs the mayor's stamp of approval. Everybody needs the mayor and his secretary." This is also true of Scorniceşti, and everywhere else. Mayors come and go, but Mrs Şerban, the daughter-in-law of the old mayor, is there to stay, and everybody knows that she is running the show. Elisabeta Rizea recounts how in the past Mrs Şerban would have a neighbor spy on her as she went out to mow the hay, to see whether she might be carrying letters or food to those in the mountains. Elisabeta says she is still afraid of her. She needed authorization for something, so she went to see her with a box of chocolates, "but my hands were trembling, just like this, I was so afraid of her." Mrs Şerban blames the situation on regime change: It was not people's fault—not her father-in-law's, nor her husband's—that they had to take possession of other people's plots. "Well, why, a law was passed which means that everything done then was legal. Times have changed, and now there is a new law, saying differently; it's as simple as that."

All the schoolmasters and teachers in Scorniceşti and the surrounding villages have been members of the Communist Party. So has the new Nucsoara priest, who replaced those who were executed. As a young man, he graduated from the seminary a long time after an accord between Party and Church had turned Orthodox priests into employees of the communist state. The collapse of communism was a blow to his own interests and to those of priests in general, as he himself admits:

> I made a better living as a priest under Ceauşescu, I did well both financially and spiritually. At the time I was paid 790 lei monthly

and the rest was my own private income, out of which I built my home. After '90 I no longer had an income from private sources, only my state salary, paid according to qualifications and seniority in work; out of that I would pay the church singer and I'd be left with a surplus. After 1991, as there was a radical change in circumstances, I no longer claimed a private income, and I received a salary from the state of fifty euros before tax, that is about thirty euros net each month. I've got land, too, I've got four hectares now, and at my mother's death I'm going to inherit one acre more. I have a cow, and I produce my own food, I don't have to buy anything for the household. I sell fruit at harvest time, that's all. And I make home-made plum brandy.

Q: And from a spiritual point of view, why do you think it's worse now than in Ceauşescu's time?

A: Church attendance has declined for lack of money. I was paid per service by parishioners. Some would buy candles in church, another source of income. Now they don't have the money and are too embarrassed to come. Before, I had a salary, and I paid the singer myself. Now there aren't enough christenings and weddings. I've got seven or eight weddings annually at most for the entire commune of Nucşoara. And because people are impoverished we no longer charge a burial fee.

[…] I'm teaching now, I teach at the seminary in Curtea de Argeş. Eighty per cent of priests in the Patriarchate are in this situation. I spoke up at a conference and asked: What's to be done? If the state doesn't take a decision about the priests, it'll be a disaster. There's been a proposal submitted to the government by the Patriarchate and the Ministry for Religious Affairs whereby one per cent of GDP could be allocated to the Church. The priests could be placed on a state salary of 100 euros and there wouldn't be any charge for the church service itself. A priest shouldn't be, as he has been and still is, a beggar, a black sheep. Let us, priests, join the ranks of intellectuals and be on state payroll.

The peasants' dependency and the destruction of the rural middle classes—traditionally, the natural patrons of the Church—has led to a situation where the priest depends on the state entirely and dreams of being on the state payroll rather than being paid on a fee-for-service basis. Like the village schoolteachers, the priests no longer represent the community and have joined the many categories aspiring to live off the state. The priest I talked to is also a local councilor, like Ungureanu and like all the other representatives of the new elite. He explains innocently that he ran as a candidate on the first list suggested to him, which happened to be the list of the former Communist Party. There is no evidence that the council members are providing any useful service to the community, unless you believe that requisitioning—for a second time—Nel Preda's lakeside land for communal fishing can count as a public-spirited project. The peasants here have no spare cash. Poverty and exemption from land tax, defended by all postcommunist governments, have therefore emptied the coffers of Nucşoara's local council. The village is paid from the state budget almost double the amount it collects in revenue. In 2000, for instance, it collected 300 million lei in tax and received 500 million from the budget. Thus they had a total budget of less than 30,000 euros, including the salaries of Mrs Şerban and of the mayor. The only public project completed in the commune was not the outcome of the councillors' efforts, but of those of Elisabeta Rizea, who persuaded the authorities in Bucharest to asphalt the road between Sboghiţeşti and Nucşoara. The councillors now regret that they did not seek to use her celebrity status to greater advantage, but they do not show any gratitude to Nicolae Noica, the minister who authorized the roadworks. Nor do they show any gratitude to Elisabeta Rizea herself.

Predatory elites

According to Barrington Moore Jr. there are three ways in which people can secure the goods and services they need. "First of all, people can make what goods they want themselves either individually or collectively. Of course people cannot 'make' admiration or distinction themselves. Other people have to grant it. But this is possible on rational rounds in a variety of collective undertakings. The other two ways for people to get what they want is to steal it, or else force other people to make it for them. [...] By definition, stealing is not legitimate and without legitimacy there is no authority. There is merely domination. [...] a truly predatory elite is one that renders very few services to the underlying population and extracts for its own purposes a surplus big enough to create poverty on a massive scale that would not exist if the subordinate population were left to its own devices."[4]

Although it forms part of the lower-level local, rather than top-echelon *nomenklatura*, the group described in the present study nevertheless has all the characteristics of a predatory elite. A village such as Scorniceşti has always been poor. In contrast, mountain villages have always been self-sufficient, and even though people have never been wealthy there, nor have they been starving. However, this changed dramatically under communism. "I've been thinking, in my own slow way, how come there was poverty in Ceauşescu's time? In ten years, they haven't found the truth about the Revolution; in my thinking, they've been really clever, 'cos it's still the Communist Party at the top," the veteran communist Gh. Şerban comments wisely. "I've been close to starvation in our beloved Romanian land," Nel Preda recalls. "I'd do swaps with the folks from the plains; they'd bring maize, I'd give them firewood. Then we were prevented from driving away from the village in our carts, and so we got to the point where we had only two kilograms of foodstuffs per quarter. In the meantime, the local party folk had a storeroom full of everything, milk, honey, caviar. And we peasants had to live on boiled nettles."

In the 1980s, the entire country suffered from food shortages of catastrophic proportions: Basic items such as oil, sugar, and bread were rationed. In the circumstances, Nucşoara was comparatively well-off because people there had livestock and, although delivery quotas were reintroduced and gradually increased in the 1980s and 1990s, they never reached the draconian levels of the Stalinist era, so people could live. Some of the peasants measure their lives in terms of the number of calves they delivered in quotas for the past thirty years, and the conclusion seems to be that the area has always been prosperous and output high. Prior to 1989 peasants could no longer find bread in their village, regardless of whether they lived in the mountains or in a plain-village, growing wheat themselves. They had to apply to the nearby town, where bakers were forbidden to sell to anyone who could not present a local ID. Nowadays, they get their freshly baked bread delivered daily by an enterprising man. During those years, all the mills, such as the one owned by Niţu, Toma's father-in-law, had been closed down and destroyed and the processing of wheat was done industrially. The peasants' access to the most important result of their work was thus politically controlled, and it produced famine. "Only the workshy who'd gotten used to doing nothing can say that communism was better," Ungureanu concludes.

Foresters and gamekeepers enjoy patronage at the highest level, given that for Romanian macho politicians hunting is a favorite entertainment. Ceauşescu himself used to hunt in the surroundings of Nucşoara, and is remembered by local gamekeepers: "He was a good shot, and a good man; once he shot a deer calf by mistake. We rushed the quarry in his direction, and he was sorry," misty-eyed gamekeeper Ion remembers. Nowadays, Ion hunts with the former political detainees, whom he addresses respectfully as "boyars." They are now allowed to carry a gun again, and Ion is not too particular about his choice of hunting partner. Neither was Ungureanu in the past, when he switched easily from Titu Jubleanu and his peers to the Securitate officers from Piteşti. Times may change, but every generation of the elite goes hunting. Today, still, the Forestry Service in Domneşti play host to important people who descend upon the forest clearing in their private helicopters, have a quick go at the game rushed within their range by the gamekeepers, and leave. One can easily squeeze it in during the lunch break: It takes only twenty minutes to travel by helicopter from Bucharest to Nucşoara. And who comes? "Our men," says Ungureanu, "Petre Roman, the former PM, Ion Ţiriac." The guests spend the night in exclusive hunting lodges, and the cars are all brand-new, glossy Western imports. Some of these lodges date back to the time of King Ferdinand of Hohenzollern and are reached at the end of alleys lined with the same old fir trees. The accommodation has been refurbished to meet new standards, as well as to disguise the devastation caused by the Securitate troops during the ten years of partisan-hunting when they were quartered there. The national head of the Hunting Association was another Social Democrat Prime Minister, Adrian Năstase. Nicu, the man who organizes hunting parties for the Italians in Scorniceşti, gets as misty-eyed about Năstase as gamekeeper Ion when he reminisces about Ceauşescu. It surely helps to have a hunting prime minister and be able to talk to him hunter to hunter. Nicu is active in the Association, casts his vote dutifully at meetings, and knows what it takes to fight those who intend to limit the rights of hunters.

In Scorniceşti things are less clear-cut, for two reasons. Firstly, the Ceauşescu family formed an important section of the predatory elite both locally, in the town itself, and in the rest of Romania. Ultimately, the Ceauşescu family had to pay for the wrongs of the communist regime as a whole, although as far as Scorniceşti is concerned their impact there was not entirely negative. Secondly, in the transition period Scorniceşti was able to redeem itself as a town rather than as a village, although it has only existed as a town since 1989 and there is still room for improvement. Towns create expanded opportunities for economic growth and social mobility, independently of the predatory elite and its degree of exclusiveness. In this particular case, opportunities were created in the textile industry, where the state no longer has control. Conversely, in areas where state control is still firm, for instance in farming or in public works, the situation has remained unchanged. The predatory elite has managed to adapt. Comrade Margine, who had imposed collectivization in one of the villages, encouraging the beatings and abuse of opponents, including his own kin, also supervised the dismantling of the collectives. In the words of one of his godsons: "How shall I put it, my godfather made and unmade the collective farm. Those were the times." In other words, the best that people can do is adapt to circumstances and go with the flow.

Likewise, comrade Turcin, Lina Bărbulescu's trusted man, became the patron of the largest farm association which sprang out of the cooperative, right down to its demise. One might wonder: Why did people elect him, after everything that he had done to them? The question is easily answered. The association has hundreds of members but, when elections are due, only the ones in the know turn up and elect the proposed candidate. "Turcin would come round to size up the courtyard and would say: 'You must be aware that the house isn't yours.' Then they'd come to measure the garden, our only resource then, those thirty units of land, called 'subsistence land'. But although it was supposed to support us, it belonged to the state nevertheless, and if *Mr* Turcin or someone else thought fit, they'd come and measure up ten times a year if needed, if that was what it took to force me to work for the collective farm, although I was a qualified engineer. Today he sits in the local council." (Intellectual, Scorniceşti)

Mayor Neacşu, although freely elected by the villagers in 1996, thought it was far better to withdraw behind the throne and work in the Finance Department, leaving the mayoral seat to an enterprising younger man. The latter was recruited by Neacşu's party—the social democrats, naturally—shortly before the elections of June 2000. Mayors come and go, but the clients of the mayor's office remain the same. The entrepreneur who built the apartment blocks during systematization—and made a hash of it—is still in charge of public works. How does he get the commissions? "He wins the bids. If he's the only one working in the area! But contracts are always fixed beforehand. If he takes part in an auction against those in Slatina, for instance, he's going to win, because he's got his own people over here, in another place the job will be given to someone who's got people there. [...] This is how it works: they arrange it among themselves, you come and counter-bid against me, and I'm going to win, next time I come and counter-bid against you, and so everyone keeps their own territories." (Focus group intellectuals, Scorniceşti) Naturally, it is all done with the connivance of the mayor's office.

One can often see members of Scorniceşti's local elite having a pint on Nicu's restaurant terrace. Often they pick him up in the car and they all go clay-target shooting in Slatina with the county notables. These are mostly the local bankers who hand out business loans, the young new mayor, and the chief of police. People think that even the restaurant business is not Nicu's and that it is merely a front. Otherwise, who could have known in Bucharest that the farming association which succeeded the Scorniceşti cooperative was bankrupt, or that the bank wanted to get rid of the hillside camping site and the adjacent "Brooklyn Bridge"? Only the local bigwigs had this information. Enter the young investor from Bucharest. He bought the camping site straight from the Bank of Religions (which in the meantime had collapsed after a scandal involving preferential loans), got a 100-year concession for water services from the council, and got the business started. The local bankers are former managers of the communist state industry. The new elite is the old elite, minus the Ceauşescu family, but plus the individuals who came to the fore with the arrival of the younger generations, people such as the mayor and Nicu himself.

These resources could be material resources such as land—as illustrated in the present study—or abstract, such as various authorizations and certificates applied for at the mayor's office, an inheritance of the overregulated communist system. The change of regime in 1989 left intact a large number of the institutional mechanisms of dependency created in fifty years of communism. This rendered the land restitution virtually redundant. The predatory elite survived the postcommunist land reforms, as they had key strategic positions for its implementation and managed to enshrine several practices controlling people's lives. Moreover, institutional change was not initiated in the countryside, but in Bucharest, where President Ion Iliescu's legislation in 1990 and 1991 encouraged the creation of farming associations and restricted the restitution of forests to private property. It was not entirely top-down though. The local predatory elites and those corporate groups that represent their interests—the communist bureaucracy, or bodies such as the Forestry Authority—sent a clear message to the center, signaling that they did not really want things to change. They became important stakeholders in the management of change. It was an insidious battle for social control, which could result in a clear victory for only one camp, as the other had been practically exterminated by then.

Notes

[1] Seton-Watson, *The Pattern of Communist Revolution*, 342.

[2] The official figure advanced by the Romanian Intelligence Service (SRI) is 115,000, but historians believe the real figure to be almost three times as high, because many of the political offences were officially registered under other categories. See Presidential Commission for the Analysis of Communist Dictatorship in Romania, *Final Report.*

[3] Seton-Watson, *The Pattern of Communist Revolution*, 145.

[4] Barrington Moore, Jr., *Injustice*, 445–6.

CHAPTER 6

The Manipulation of Lifestyles

On 19 January 1989 the prestigious fortnightly *The New York Review of Books* published an open letter by three academics from Princeton University under the gloomy title "Razing Romania." The three historians, Istvan Deak, Natalie Zemon Davis, and Carl E. Schorske, disclosed the plans of Nicolae Ceauşescu to raze half of Romania's villages by the year 2000, starting with the smaller settlements with immediate effect. After ten years in which there had been much talk with little action, the program of "systematization" had at last started triumphantly in March 1988. Information had reached the three authors about the pilot sites in the suburbs of Bucharest, the villages located around Otopeni International Airport, whose fate offered a clear prospect of what was in store. So far, systematization had resulted in demolishing the modest peasants' houses and cramming the displaced residents into improvised collective housing projects. In Bucharest itself, the demolition of the historic downtown, including some heritage-listed churches, was under way. "The ancient Saxon-German, Hungarian, and Romanian churches and cemeteries in the villages are now threatened with destruction beyond resuscitation. [...] Their capricious destruction would diminish the whole of the European cultural heritage. [...] Readers in the United States can write to the Embassy of the Socialist Republic of Romania, 1607 23rd Street NW, Washington, DC 20008," the authors advised.

By the time the readers of *The New York Review of Books* would have had a chance to write to the embassy, and long after several West European states and organizations had expressed their concern, in Scorniceşti systematization had already been completed and the project was already being extended to the commune's other villages. Ceauşescu's fellow-villagers benefited from the effects of rural modernization before the other Romanian peasants. The real impact of this intervention can only today be assessed.

Enforcing systematization and urbanization

Systematization was first mentioned at the 10th Congress of the Romanian Communist Party in August 1969. The debate returned in greater detail to the Party's National Conference in July 1972, and, finally, the Central Committee's plenum of March 1974 endorsed the draft law on the systematization of the territory and localities of the Socialist Republic of Romania. The Popular Council of Olt county was way ahead: As early as 1973 it had approved plans to systematize the county's so-called "reconstruction" areas.

In Scorniceşti, systematization started in 1970, after the commune had already expanded as a result of administrative–territorial reorganization. A new village hall and two or three housing blocks had been added in quick succession in a bid to upgrade the village, but it still did not live up to the new ambitions of its promoters. It could hardly compare with Negreşti, which had now become its subordinate village. The ambitions of the Ceauşescu family, and especially of the local branch, the Bărbulescu husband and wife, led to a large-scale redesigning of Scorniceşti, which drew on the party's general blueprint of a cluster of collective housing projects built around a "civic" center whereby the Culture House replaced the church as the symbolically charged focal point of the community. Initially, it was nothing more than an urban development project: The new civic center, around which the town itself would have later developed, was to be located between the bridge and the European motorway, practically on the edge of the village. The project was awarded to Proiect Bucureşti, the country's leading architectural institute, previously responsible for the demolition of Bucharest's historic center to make room for the projected "Avenue of Socialist Victory." But Ceauşescu rejected the plans, which would have turned the road to Slatina into a near-highway. He said: "Don't touch the road I used to take as a kid on my way to school." Witnesses to this scene claim that these were the only words he said. The people who allegedly had a monopoly on understanding Ceauşescu, the Bărbulescu couple, both now dead, interpreted these words in a way which was to radically affect the fate of the village. Plans to build the community center outside the village perimeter were scrapped, and it was decided instead to build it in the area between the two bridges, exactly on the site of the old village. The road was widened and asphalted, but otherwise it remained untouched. Properties were expropriated by decree, and demolition started, with "villas" and blocks of flats being built to rehouse the residents. The villas were an emblem of local privilege; other places throughout the country had only housing blocks on offer. The so-called villas were in fact houses with two or three storeys and two entrances, which would accommodate two families and allow for a strip of garden, preserving some of the peasants' lifestyle. In the words of the then mayor:

> In Scorniceşti systematization started in the same way as everywhere else in the country. The objectives were set and approved in 1975 and, as Scorniceşti aimed to become a powerful agro-industrial center, a feasibility study was conducted by Proiect Carpaţi and the planning section of Olt county, as well as by specialists from the former council of Olt county.
>
> For the period '80–'85—the five-year plan—it was forecast that between 6,000 and 8,000 apartments would be built. Under Ceauşescu's personal supervision, this figure was revised downwards from 5,800 to 1,600–1,800 apartments. Demolition work started around 1975–1976. From 1976 to 1979, when I was elected mayor, only three houses had been demolished. Let me tell you, here in Scorniceşti we had eleven mayors in ten years. It was a difficult time. This was a disaster area: a six-meter-wide road, no pavement, a ditch on both sides. If a car crashed into one of those ditches, there was no way of lifting it out, not even with the help of a tank. The villas had been built in such a way that the village was left on one side, while the town developed in the opposite direction. When Ceauşescu said "Don't touch the road," they started moving the town in the opposite direction of the village hall. Soon, people realized that building was heading in this direction and so they would lose their gardens and courtyards. They ended up in apartment blocks. In Scorniceşti it was decided to build three-bedroom apartments and in the end these far outnumbered the others. Why? Because the intention was to have parents and children live together, in the same way they had formerly lived in houses, with the entire family, father, mother, daughter, son-in-law, and so on. In fact, that wasn't strictly accurate, only the yard had been common; they used to live in separate sheds, otherwise they'd be at each other's throats all the time! So, when they were reunited in the same apartment, scandals started. Families even

> avoided moving to villas, because in a block they'd be able to live separately, whereas in villas they'd have to share a courtyard. So, if three families shared the same yard, they were sent to live in three different apartments.

A total of 178 properties were demolished in central Scorniceşti, and slightly above 300 overall. The former mayor, Neacşu, says that the demolished buildings had mostly been made of timber or half timber and only three of brick, so the loss was not great. However, the layout of a peasant homestead, even a modest one, differed greatly from that of an apartment. A typical peasant house would have two rooms, one in which people stored their good clothes and the women's dowry chests, and the other which served as a bedroom. Many houses would have a lateral extension too, used for storage. The houses also had terraces and porches, flowerbeds at the front, a courtyard at the side, and a garden at the back. The courtyard accommodated the stables, the chicken pens, the kitchen, and, further back, the lavatory. Among the first to move out with their entire families in order to set an example to the community were the "cadres." One of them, the director of the Culture House, declared that country life had never suited him anyway. But few others were happy to move. Ioana Mardare and her husband had had a brand-new house built in the early 1970s.[1] Lică Bărbulescu had often warned the husband to stop investing in the house, because it was going to be pulled down, but the man was convinced that only the old, dilapidated buildings would be demolished. However, Lică had told him explicitly that it was all going to be done "in due order," house by house. The Mardare family did indeed duly receive an expropriation order and were allocated an apartment in one of the newly built blocks. The demolition men's greatest satisfaction was that the first property to be demolished belonged to Nana Iana, practically the only person in old Scorniceşti to have stayed out of the collective. The first building to be condemned—"disaffected" in the period's jargon—to make way for the textile factory and the "machine tool and assembly unit enterprise" was that old woman's house, complete with stables and chicken sheds. The former mayor recalls cautiously: "I was not directly involved, but because she was the first, and a straightforward, brave woman too, she caused a lot of trouble." The man in charge of systematization at the time, Mr Pătraşcu, is today still the chief urban planner for the council. He recalls how they went about persuading people:

> *How did people react when you started demolition work? Did you meet with any resistance?*
>
> Yes, it was very difficult. Don't get me started on that! Just to give you one example, the single residents' hostel, a building not far from here. It was in 1975–1976, just after the first expropriation decree. A number of houses were to be demolished to make way for the textile factory and a workshop, such was the plan, to be followed by a hostel for single residents. We were supposed to help with the demolition, dismantle the fences, this and that.
>
> *How many homes were due to be demolished?*
>
> Oh, some five or six houses. So, we made a start. This man had a very well-looked-after vineyard and when we told him we had to remove everything, down to the timber posts, he just snapped, grabbed a hatchet; the deputy ran off, I had no clue what to do, but some people managed to take hold of his legs and he fell before he had a chance to hit me. I'll never forget that. In the end he got a grip on himself, poor chap; he was desperate.
>
> *And did you notice any sign of eager anticipation in anyone? After all, they were moving into new, well-equipped homes.*
>
> People would have been quite happy to be allowed to stay on in their own homes. After so many years, they'd got used to the way they'd lived in their old homes.

Aside from the fact that people had to abandon their old parental homes, which they had been improving all their lives, leaving behind their gardens and livestock was a blow to the peasant economy and lifestyle. Nana Iana never gave up her animals, although she did say she had to "slaughter some of the goats." She built a shed in a sheltered spot not far from the new building to which she was relocated, where the goats' presence "did not affect the cordial coexistence," as the communist mayor said in his inimitable style. She did, however, "keep some of the goats under the balcony, from where they would sometimes escape into the future town's parks, where they'd graze on the flowerbeds," the mayor complained. When Ceauşescu's visits were announced, there would be mass mobilizations of people who then had to round up the goats and the geese, remove them from the new boulevard, and keep them concealed at least temporarily. After 1989, the animals too won their freedom, grazing on the flowers adorning the central boulevard, a favorite meeting point for dogs turned stray due to the demolitions.

People received some financial compensation for their demolished homes to pay for new housing. The land, however, having been expropriated, was lost. Out of their compensation money, people would first pay a thirty per cent deposit on the villa or flat where they were going to live and then pay off the rest in modest installments over fifteen years. However, the new houses were more expensive than their old homes, so they ended up in debt; villas were also more expensive than the apartments. The plots surrounding the villas—around 400 square meters each—had to be paid for separately, a disincentive for potential purchasers. People, therefore, preferred living in the housing blocks, where at least they had running water. Not that they knew how to use it. The mayor remembers the crusade to civilize the villagers: "It was tragic, really tragic. One had to show them how to flush the toilet, and tell them: Look, that's the water tap, you have to turn it off, otherwise you'll flood the house. The new trend in those days was to have a little doormat at the entrance to the building, instead of floorboards. Well, they'd take their shoes off as they entered the building, long before getting to their apartment; you'd never believe it! One of them broke his water closet as he was trying to figure out how it worked, and then he shouted: 'Alright, I've got it!'"

The sudden, mandatory change turned life in Scorniceşti upside down. Additionally, in the same period, the town was being industrialized and had to import a specialized workforce, which also had to be accommodated. The puzzling thing is why it was necessary to condemn existing houses, when there was so much available space in the surrounding fields. Blocks could have been built solely for the new arrivals, and that would have left enough space for the must-have civic center too. But farmland was legally untouchable, so the legal and administrative red tape led to "building condemnation" instead. But why was there a need to protect farmland, when Romania, if anything, had a surplus? It was due to the renewed stress on increasing agricultural production and ensuring "continual growth," a concept reflected in the annual plans providing for increasingly greater levels of output. The party daily *The Flame* of 15 July 1982 reported that from 1950 agricultural output had increased more than threefold. "In their determination to surpass themselves and secure ever better annual results, in 1979 the members of CAP Scorniceşti launched a challenge to all the other agricultural cooperatives in the country. 'It is not easy to become a front runner,' Dr Vasile Bărbulescu, the cooperative chairman and a Hero of Socialist Labor, said at the time, 'but to stay a front runner is even harder.'"[2]

Given the strong membership of CAP Scorniceşti, the condemnation of properties was not a foregone conclusion, but making an exception would have required political will. After all, the Communist Party still believed that peasants were essentially, in Lenin's words, a counterrevolutionary, if inert, force. The processes of modernization were meant to suppress the peasantry and turn it into a more advanced class. Therefore, the annihilation of the peasantry by effecting a radical change in its lifestyle was at least as important an objective as high rates of urbanization and agricultural output. The disappearance of a backward village culture meant the deletion of its collective memory, and a brand-new one could then be constructed from scratch.

> The decrees of mass expropriations were issued only for high-concentration areas [the mayor explains], and just one year before construction was due to start, because building was done at a slow pace. After all, one had to be able to offer apartments to all the newcomers. Well, out of the fourteen villages that belonged to the commune of Scorniceşti, only ten had areas apt to be "reconstructed." So what were "reconstruction areas" and how were they established? There were criteria, such as birth rate, mortality, etc. And it was all about high concentration: you were no longer allowed to build in the traditional way, houses with yards of 1,000 or 3,000 square metres. This was in order to gain land for agriculture. The emphasis was on brick buildings with all the amenities, in areas with their own schools, kindergartens, medical practices, pharmacies, and so on, all within a narrow perimeter.
>
> This was very time consuming. In everyday life, it meant that if a particular village was marked out for gradual condemnation, no new

> building permits were issued; they would just give you the alternative in the new building area. Or, if the lamppost in front of your house was defective, they wouldn't repair it, and so gradually people were forced to leave that area. This was one stage in the process. We didn't have enough money to have an entire village demolished, for there would be 112 households and you'd have to give everyone alternative accommodation. And we had difficulties with the industry too; new people were coming in all the time. Some 1,060 cadres arrived in one year alone. We had to give them all apartments. There would be engineers hired by the Pulsor factory, and by the brewery, and there was a fast-track accommodation scheme for those who came by government appointment. Building started wherever there was a vacant site. In the area next to what was the Ceauşescu memorial house, all the properties were built on vacant sites. In some cases, only half of the house was demolished and the new villa was built right next to it; and so they stayed there, with the villa tucked in against the rest of the house.
>
> Thirty years from now, there'll still be livestock here in the city center in the green spaces between the blocks.

The peasants had no means to oppose the demolitions. Opposition to systematization in the 1970s and 1980s was quiet compared with what had happened during collectivization. In the intervening decades, the state had significantly increased its control over society. Many people had become state employees, either in industries or on the mechanized farms, and could be threatened with dismissal. The peasants now had state pensions and, because they no longer had land, depended largely on those pensions. No alterations to properties were allowed without a permit from the council. Pressure on the peasants could, therefore, be exerted in many ways. Their autonomy had received the *coup de grâce* with the disappearance of their gardens, which were supposed to produce the family food. Producing food in one's own private garden had become a legal requirement in 1981 as a response to the food shortages. Had the dismantling of the private gardens been extended to the entire country, and not limited to Scorniceşti, there would probably have been mass starvation. Special privileges were granted to transfer foodstuffs from Slatina to Scorniceşti. But this did not solve the problem, as people could not be expected to adapt overnight to life in the blocks of flats. They took their earthenware pots and their wooden spoons and moved them into the modern apartments. These were more spacious than the traditional peasant homes and had all the basic mod cons, although they had been built in a rush and had many hidden flaws, which became apparent only later. The new residents continued to keep their livestock in the areas between the blocks. Building was usually done at a mad pace, but only in fits and starts over the years. Once a visit by Ceauşescu was announced, there would be a quick outburst of window-dressing in which façades would be built in two weeks. This was followed by longer periods in which attempts were made to respond to the planning errors of such massive development, which could, however, no longer be altered. Once authorized, neither the planning nor the design of such projects were open for negotiation, and the builders struggled to tackle problems of detail, or carried on regardless.

A new infrastructure had to be created overnight. It was completed, with many flaws, in ten years. A system conceived for sixty housing blocks had to be adapted to 115. The water and sewage services were particularly problematic. Problems with the current infrastructure sprang directly from the initial faulty planning conceived during systematization. Older plans, for instance, would be used unchanged for larger structures. Or water pipes would be put under the asphalted road, which created problems every time repairs were needed. The water wells were dug very deep, because there was not enough surface water to cater for such a large human settlement. But these required high maintenance: today, out of eighteen wells, only six are still operating; the rest were already clogged up by 1989. There are no funds for unblocking them, because funds are no longer being diverted from other localities to Scorniceşti. Initially, central heating had not been planned for Scorniceşti, and heating was to be provided using local materials such as wood and coal. There are no natural gas resources in the area. Initially, the apartment blocks were to be equipped with firewood stoves. Subsequently, the number of flats was multiplied to achieve a higher density in a smaller space and the new plan envisaged the introduction of central heating and long-distance heating. A large heating station was built in cooperation with the Pulsor engineering firm.

Building went on uninterrupted up to 1989. The entire project was still unfinished at the time of the Revolution. The next stage would have involved the construction of administrative offices, a hotel, a community arts center, and a plaza with a small park and a fountain. The only things to be actually built were an avenue lined by two rows of apartment blocks, another cluster of blocks at some distance, and a few villas. However, Lică Bărbulescu achieved his great ambition: the building of the great stadium, which used up most of the funding, so that the civic center was put on hold. So Scorniceşti, which did not acquire civic status until 1989, was left without its civic center.

After Scorniceşti, it was the turn of the neighboring villages to be reconstructed. In their case, however, the authorities did not have to tiptoe around local sensitivities. Nicolae Popa recalls those times:

> Once they'd completed the center there, Mrs Bărbulescu got the idea that they should start demolitions in our village Roşiori as well, by the lakeside near the border of Scorniceşti. [...] Instead of the village, she wanted a vegetable garden, just like that. I was working at the council at the time, so I learned about this plan to have the village Roşiori demolished. I was a councilor for five legislatures, here locally, as well as in Spinei.
>
> *What did it mean to be a local councilor at the time?*
>
> Don't you know what it means to be a councilor? I represented the will of the villagers, of those who had elected me.
>
> *Do you mean you represented their will politically?*
>
> Administratively. So I learned about this plan and I said, wait a little, I'm going to write a memo and send it to my superiors, to the president of the local council. So, I wrote down: "I wish to submit the following proposal. I would like you to reconsider the possibility of making changes to the future urban centre of Scorniceşti. The two villages, Roşiori and Teiuş, should be joined into a single village under the name Teiuş, within a reconstruction zone encompassing the areas currently populated. I am making this suggestion because I believe that it would not affect in the slightest the future interests of Scorniceşti, which is an agro-industrial town. There are 726 active persons in the two villages currently, and 224 households, covering large areas of high-quality arable land. Given that there are two reservoirs in the area, the irrigation facilities can considerably help the expansion of farming and animal husbandry. This plan should pose no financial difficulties, because the roads have already been asphalted, and both

villages have got power lines as well as a new school, a playgroup, and a shop. Historically, the two villages have existed for hundreds of years as a single commune. This arrangement lasted for a long time. Meanwhile the inhabitants fought in the Independence War and in the two world wars. They were in the frontline promoting the policies of the Romanian Communist Party soon after 23 August 1944. Twenty years ago, they formed a single collective farm which contributed four hectares of arable land per household to endow the new CAP Scorniceşti. Currently, the locals are front runners in all the activities and projects run by the CAP Scorniceşti. Comrade President, I sincerely hope that you will forward this application to the county commission and that this village will duly be included on the map of the new urban centre."

And were you successful?

I was indeed, and the request was forwarded, and one day there comes this panel from the county and they ask to see me, comrade Popa. And they said: "Could you come and show us around?" It was the president of the commission in person. He surveyed the village and said "Look, you seem to be a reliable chap. You're totally right, but we can't do this, because Lică Bărbulescu is opposed."

So did they want to demolish you there and then, in 1989?

They wanted to destroy us. Oh! Glad you reminded me. Two or three days earlier, these two tractors and bulldozers pulled up in front of my garden, and the drivers said: "There, we've come to take out your vines and the trees." They said they're going to start demolishing the village.

So you were a kind of example for the rest of the village?

They were afraid that others might rebel. And I said "C'mon, lads, not so fast. Leave us a while yet, 'cos we have stuff here, trees and vine posts. Just give us two days to remove all this and then you can come and do as you please." Well, they say, "We have orders from papa Turcin." This is what they called him, papa Turcin. He had only to say one word, and they'd come and pull your house down.

Did Turcin have anything personal against you?

No, he didn't, but he was in charge of the demolitions.

But why was he in charge? He was the president of the cooperative farm.

Well, because the freed land would have gone to the cooperative afterwards. It was the head of the CAP who supervised the demolitions, here in the countryside, not in the town center. And as luck would have it, I spotted this engineer here by the lake, the guy who came along with the tractors to pull out the trees and the vines and the rest. [...] So I say to him, "Look, Mr Engineer, they've come to kick us out. Could you please ask these men to kindly go home and come back after the weekend, on Monday, and give us time to remove some trees and the vine posts, and whatever, and then we'll go?" And he did send them home and told them to come back Monday morning. And we and the neighbors got down to work, pulling out the trees and everything. In the evening, as we listened to Radio Free Europe…

Were you listening to Radio Free Europe at the time?

Yes, regularly. And so, as we listened, we heard about what had happened in Timişoara, where people rebelled. In great detail, everything, hour by hour as the events unfolded. When I got up next morning, the bulldozers were gone; they took them away at night and we never saw them again.

The villages of Roşiori and Teiuş were being threatened with "gradual condemnation," because they each had fewer than 500 residents, a size deemed to be non-economic. The village of Piscani in the same commune was lined up for "rapid condemnation," which meant total demolition. Nicolae Popa was trying to stop the destruction from within the system. He had not just Lică Bărbulescu's opposition to fight, but the Communist Party's entire program. Most of Romania's 13,000 villages had fewer than 1,000 inhabitants, and many had fewer than 500. But the size deemed ideal for the "rationalization" of public services in the countryside was 3,000. Those villages which fell short of that figure were going to be listed as non-viable. Nearly 3,000 such villages were going to be condemned gradually, the rest were to be brutally demolished and relocated. Piscani and Nucşoara were in the latter category, as they were both very small. Nucşoara was lined up for demolition in 1990, and its residents were to be relocated to Mioveni, a settlement which had evolved around a nuclear plant.

Because of its privileged status, Scorniceşti had been grouped with the approximately 550 villages due to receive funds from the urbanization program and be turned into agro-industrial towns. But their growth as towns was slower than expected. At the same time, the village demolition program was stepped up starting in 1988, when the number of villages to be destroyed was increased from 3,000 to nearly 8,000. By then, however, the days of the regime were already numbered. Any delay in its demise would have had tragic consequences for many Romanian peasants. Had the Revolution occurred later than December 1989, Roşiori-Scorniceşti and Nucşoara would not exist today.

The everyday life of the agro-industrial town

Scorniceşti was obviously used as a guinea-pig. It was the object of an experiment which tested the feasibility of the agro-industrial town, with fewer than 20,000 inhabitants, designed to provide services to the agrarian areas surrounding it and sustain itself mainly through agriculture. The choice of Scorniceşti was justified officially by reason of its better soil and short distance to the nearest town. But the transformation of this mega-commune into a town was not an isolated act; it was part of a national strategy aiming to reduce migration to the big cities, closed following their flooding by rural migrants during the industrialization of the 1970s. A similar project targeted the commune of Domneşti, not far from Nucşoara. Surrounded by coal mines, this commune was believed to have the potential to become a town. But this ambitious project did not benefit from the level of funding which went to Scorniceşti and, consequently, there were few changes in Domneşti.

Demolitions and the consequent disappearance of yards and gardens changed the countryside fundamentally, but there were subtler measures with an equally devastating effect. Firstly, cooperative farming promoted a collectivistic organization at the expense of the individual enterprise. Secondly, the coexistence of rural and urban lifestyles in the same area had an impact on attitudes and behaviors. By the 1970s, a large number of households had a mixed structure, in other words at least one adult per household worked in a town and earned a salary. "Then these factories appeared, so people got jobs there, they preferred working there. They were better off, and the employment card gave them some security. They'd no longer go out to work in the field. These were people who were born peasants, yet they'd no longer work their land now that they'd got these jobs in the new industry." (Focus group graduates, Scorniceşti) The individual plot supplied the necessary food. If you add to this the pensions, salaries, and child benefit—quite generous because of the regime's pronatalist policies—and it becomes obvious that incentives to work for the cooperative were declining. The regime itself was quite willing to turn a blind eye, because cooperatives did not produce enough to give members their yearly dividends in kind, as their contracts were stipulated. The peasants, in turn, no longer tended the common fields and preferred working on their own plots instead. A survey conducted in the 1980s but never published showed that one-third of the cooperative members had not put in a single day's work for the farm during the year.

> We didn't get enough produce from the CAP, which is why all those who could were stealing. The Bărbulescus were not as bad as people say. They knew that everybody was *at it*, but they'd turn a blind eye; they never sent anyone to court, not even when they caught them red-handed. (Peasant woman, Constantineşti)

> Everybody was stealing. They'd pass under my office windows at the council, and I'd say to them, "C'mon, man, take the trouble to go round that fence at the back, rather than do it under my nose; they're going to fire me one of these days." (Nicolae Popa)

> They were all stealing like crazy. The agronomists would steal, the school administrators, staff at various institutions, and of course the peasants, too, when they saw that everybody was at it. (Teacher, Negreni)

Theft at the cooperative in Scorniceşti became as widespread as wood looting in the forests near Nucşoara. One might wonder why under the cooperative system such misdemeanors were largely tolerated in the eighties, whereas previously people would be sent to the gulag even for imaginary crimes. The reason was that by this stage the regime had already achieved its main objectives: it had consolidated its power while depriving the peasants of their autonomy. The regime was aware that people were unable to make ends meet with their official gains, and it tolerated theft to help its welfare system, and so relieve the growing social discontent. Stalin had wisely warned against idealizing the kolkhoz: "It is only a weapon, it is nothing in itself, it is an instrument, not an ideal." Theft was tolerated because it, too, was an instrument in the hands of the regime, to the extent that it made everybody conveniently vulnerable.

It was a surreal system, growing stranger every day. Fertilizer would be pumped into the soil, producing huge crops of 6,000–7,000 tons per hectare, whereas after 1989 a hectare could barely produce 5,000. Experimental plots yielded up to 17,000 tons. Popa was responsible for forwarding the results once they had been centralized. If party documents said that one plot had yielded 17,000 tons, the Party daily *The Flame* put the figure at nearer 25,000. The agricultural land in Scorniceşti proper, where peasants had been poor, was smaller than in Mogoşeşti, where individual plots above five hectares were frequent before collectivization. In the 1980s, the 200 hectares of Scorniceşti came to be tended by local employees who worked mandatorily, threatened with the loss of their jobs, and by secondary school students during their school days. The latter worked for a month in the autumn and another in the spring, while employees were called out mainly for digging and raking. "But the pupils were paid, so they were more interested. They'd be paid 80 lei. And teaching staff the same; they'd get 120–150 lei, but we worked from morning till night. However, given our modest salaries, 1,500 lei didn't come amiss." (Teachers, Scorniceşti) It may not have been much, but the village intellectuals could not refuse. As state employees, they were even more vulnerable than the peasants. As one engineer narrated, comrade Turcin and his staff would come to measure their gardens, as these "subsistence plots" were in fact state, not private property, implying he could confiscate it anytime so as to force people to work for the cooperative. The threat was enough.

> The main workforce was the pupils, and others only as an addition. Because you could call out over a thousand pupils at any time, but you couldn't call out a thousand peasants when needed. So state employees would just fill in and do whatever was left to do, working alongside the few cooperative members. It was cost-effective, because the cooperative got a sizeable workforce out to do the work,

> yet because they were students they didn't have to pay them that much. Students from the Agricultural College were easier to mobilize and were paid [...] and they did not need the yearly dividends in kind that peasants required." (Teaching staff, Scorniceşti)

> We in Călineşti felt offended when they'd call us out to work the fields and we had to work a month. But in Scorniceşti they had to put in two months of work per year. I know cases where they went hungry, because the people supervising them, the college administrators, would steal the food rations. There was no way you could refuse to go. We teachers would lose our jobs if we refused. (Teacher, Negreni)

Although the village had been officially upgraded to a town, such coercive methods as confiscation of individual garden plots served to keep state employees stuck in their former peasant condition, even when they made efforts to emancipate themselves. The cooperative in Scorniceşti had the best rates of pay in the county to motivate peasants to go out and work the land. However, at this stage money was no longer a sufficient incentive. Younger people had taken up industrial jobs in town, in Slatina, Piteşti, or Bucharest; the middle-aged had found employment locally; and the elderly were happy to potter around in their gardens, and could no longer be bullied to work at the collective farm. The system worked infallibly only when it needed to blackmail the new settlers employed in local industry, for instance. They could be laid off or kicked out of the party, and their families depended on the state for education, insurance, allocations, and mortgages, and had no gardens to live off. But the system was powerless when confronted with someone like Nana Iana. They had already demolished her house and stables. What else could they do to her when she released her goats into what the communist mayor called quaintly the "future town" or refused to work even a single day at the collective farm? Nothing at all.

The Scorniceşti of the 1980s was a paradoxical place indeed. Last-generation machines were imported for the new textile factory, installed in the stadium ring, but there was still no running water. High-caliber university graduates received mandatory jobs in Scorniceşti through the national job allocation program; some chose the place themselves, thinking that it might be a good career move. And it was not enough for Scorniceşti to have the only agro-industrial college in the region, it also had to have a new hospital with all the specialist departments. Physicians had to be brought over from Bucharest and Slatina. But working as a doctor there was not too bad. Those were times of major shortages, yet supplies of medicine were better than elsewhere in Scorniceşti. The Pulsor factory was built starting from a section relocated with its entire staff from the Dacia car-making factory in Colibaşi. By and large, the young intellectuals who got jobs in Scorniceşti after graduation were quite happy, especially as they were fast-tracked to houses and apartments. On the downside, teachers were forcibly recruited to work in the fields. So were the engineers, though less often. It was at that time that the current mayor—an engineer working in light industry—arrived in the commune. So did the main leader of the opposition and his chief adversary. Both, now nearing forty, are businessmen. Under Ceauşescu they were both Communist Party members: If the party secretary suggested it, an up-and-coming young man had to join, otherwise he was left with a tarnished political record. Many of today's local industrial managers arrived in Scorniceşti in the same period. Marlene, the woman who runs what is now the largest private textile company—which employs 800 employees in Scorniceşti alone—used to create designs for the state textile factory together with her husband. But she did not find that convenient, and chose to take up work independently within the Socialist Retail Cooperative Network, a niche the communist state allowed to the independent activities of small-scale manufacturers, thereby allowing it to control them at the same time. Today she and the cooperative network's 600 employees own the company and have orders from many foreign companies, including Marks & Spencer.

The 1980s also witnessed the vigorous promotion of Ceauşescu's pronatalist policies, which made abortion illegal unless a mother had had four children already. Contraceptive methods were primitive both in urban milieus and in the countryside, and abortion had been the general method of birth control prior to this policy. Pushing for more children in times of food and energy shortages was bound to produce more suffering, yet some took advantage of the incentives offered by Ceauşescu's policies. The local pediatrician recalls:

> Gypsies were having babies, because they got state support for every child, so if they had many children they'd be able to get some cash,

> as they didn't have jobs. For instance, they'd get baby milk powder, of which I had loads and didn't know what to do with the surplus. I remember I had to give away milk powder, a kilogram and a half per child per month, but ordinary families wouldn't have it, they preferred imported infant milk. I had to somehow get rid of it, so I'd give away milk powder so it could be used for puppies and piglets. No, honestly, I had to get rid of it, because *I'd get criticized if I had stocks of it.* So I was in a situation where I had plenty of milk powder boxes left and nobody to give it to. I gave it as supplementary allocations, I committed all sorts of irregularities, I changed the names of the recipients, I had to find ways. It ended up with the Gypsies, because they alone had a use for it. It even discouraged them from seeking jobs, as they could make a good living just by selling it on.

Running water was lacking, but people could have bathed in huge lakes created from the infant milk powder. As for the other kind of powder, the dust that rose in clouds from the demolition and building sites, there was no shortage of that either. Indeed, there is still plenty of it today in Scorniceşti.

Modernization and demodernization of the village

In the 1980s, Scorniceşti enjoyed better treatment than the rest of the country during the severe food rationing, but suffered as a consequence of the coerced modernization plans. The change heralded by the Revolution of December 1989 came just a few months after the village was declared a town. That was the moment of truth for Scorniceşti. Its unsustainable institutions, such as the football club, vanished overnight. The slaughterhouse and greenhouse disappeared following more complex problems: the lack of market, the corruption of would-be managers who sold them off piece by piece, and the accumulated debt. Pulsor, the tools factory, managed to survive a few years, due to its monopoly of the production of pumps which they made for Dacia. But Dacia was bought by Renault and the French firm no longer ordered its pumps from Scorniceşti. Pulsor found it difficult to find other markets. Out of the 1,330 employees the factory had in 1989, fewer than 200 found jobs elsewhere, retired, or left through natural wastage. Some 170 chose voluntary redundancy. By 2002 676 employees had become shareholders by the time the factory was privatized in a management and employee buyout (MEBO). Consequently, its chief managerial objective was to secure the personnel it had. In 1997, as part of a macroeconomic stabilization package, public sector employees in non-viable industries were given generous severance payments to encourage them to quit voluntarily. "When they were made redundant they were paid salaries for one year, went away, bought a cow or a goat and started farming. Some started up businesses," one engineer recalls. However, many later tried to return to Pulsor, but failed to get rehired.

The privatization of a company's own employees, a method known as MEBO, was also used at the textile factory. There, it was even less successful. The workers, used to equal pay, concluded that management were earning too much. There was a revolt and the manager—today the mayor—was forced to go. He immediately set up a private company and took away many of his former clients from the state-owned firm. The prolonged strikes drove away investors and customers alike from the state factory. During the strikes, the manager had sent orders out to small workshops, rather than leave them unfulfilled. Those small improvised workshops later became fully fledged factories. Marlene herself started off with a client who saw two of his supply vehicles sequestered during the strikes. Once he had managed to retrieve them, he never returned to the state factory. Today, the former state factory is on the brink of bankruptcy, while the smaller textile factories are thriving: There are now fourteen of them, and they offer employment to large sections of the town's female population and to the young. The brewery was privatized under a different scheme and it too is thriving: there will always be a demand for beer. The former tractor station (SMT), a company which used to provide farming machinery, was also privatized under a MEBO scheme. It also tried to form its own farming association, helped by its monopoly of the farming machines in the early days after the 1989 Revolution, but failed. Today, there is not a single window-pane in the huge garage which used to accommodate Scornicești's once-proud tractor fleet which has not been smashed.

You do not need statistics to see that state industries based in Scornicești were ruined by collectivism. With employees also being shareholders, managers' lives are complicated. "It's difficult, very difficult," the director of Pulsor complains. "The annual general meeting is a nightmare. The main problem is insecurity over what tomorrow may bring, fears over job security and pay; everything that happens in the country, the recession and everything, makes people nervous wrecks. And even people who've had good, secure salaries, can now sense, like those animals released back into the jungle after being in captivity all their lives, that there's a storm ahead and things are going to get tough and they are not fit. And no matter how good a leader and a manager you may be, in a general meeting you can suddenly find yourself out of a job; everybody wants salaries of over 100 euros. Now, how can I give them that when the whole economy keeps shrinking?"

However, the investment in well-trained, adaptable, inventive personnel has paid off and when growth resumed after 2000 Scorniceşti had some companies doing well. The new entrepreneurs of capitalism saved the communist town and prevented its decay back into the status of a village. But is "saved" the right word? There is a lack of money to improve public utilities, water, sewage, and heating, although the mayor is hopeful. Textile employees earn between sixty and eighty euros a month, less than the national average. Everybody works under the "Lohn system," a form of international outsourcing in which design and raw materials come from a foreign supplier who then imports the finished goods. The local factory has no air-conditioning, so during the hot summers in the plains working conditions are really oppressive. But if workers complain, they are out of a job. They are blissfully unaware that workshops such as theirs are called "sweatshops" and that Western students organize protests to have similar outfits in Asia closed down. A draft law was introduced in Romania to ban the Lohn system. But that would mean ruin for the town. At the same time, workers in the private and public sectors are not necessarily on the same wavelength: State employees look down on their counterparts in the private sector and call them "slaves." Those in the private sector counter-attack, saying that they are happy with their lot and state industries are doomed anyway.

After 1989, many banks opened branches nationwide. Some, such as BankCoop, went bankrupt. Others, such as the Romanian Bank for Development (BRD) and the Romanian Commercial Bank (BCR) are thriving, counting both firms and private investors among their clients. Over 1,600 locals from Scorniceşti opened accounts with the

BCR, which has since been bought by the Austrian Erste Bank. The 216 private companies and 199 family associations and independent manufacturers are also with the BCR. The private sector is responsible for sending some two million dollars rolling through the banks every month. Of the 8,000 active individuals in Scorniceşti, only 268 are claiming jobseekers' allowance, which is only 3 per cent against a national average of 10 per cent (Bucharest itself has a 4.5 per cent rate of unemployment). The textile industry employs 2,700 people.

Public services have been deteriorating due to design flaws and inadequate maintenance. The deep-level water wells are equipped with electric engines which break down frequently or stop working because the electricity company cuts off the power for non-payment of debts. The sewage system was built far too close to the domestic water supply network. Consequently, it flooded over, infesting the drinking water; basements are literally full of shit. Scorniceşti's urban infrastructure collapsed as soon as the postcommunist state stopped subsidizing the domestic energy supply and prices were deregulated. People could no longer afford the far too expensive energy, and Pulsor could no longer cover domestic bills too, having its own difficulties. The central energy supply station had to be dismantled, and smaller units sold off. The residents "plugged off," the new expression used in the transition period to mean that apartments voluntarily switched off the common energy supply. Like all other residents, Ioana Mardare had a wood-stove installed with a chimney pot through the apartment window. The main difficulty is storing the firewood: Where does one keep it without a yard? So firewood has to be carried home daily in small amounts. In winter, in Scorniceşti one sees people carrying firewood to their flats in shopping plastic bags, from quite afar because the forest has also been cleared to free land for farming. A few hundred comparatively prosperous people have been able to put in central heating for individual apartments.

In 1989, the birth rate in Scorniceşti was thirteen births per thousand inhabitants; today it is just eight. General mortality has increased, and the town is in third place for TB in a country which has the highest number of TB cases in Europe. The hospital had better days in the distant past. Many specialists left as soon as they could after 1989. Those who stayed have to cope with poverty, long commuting, lack of education among the local population, and alcoholism. In the summer there are outbreaks of enterocolitis due to the infested water. Chronic diseases are on the increase. Peasants from the commune's more marginal villages do not come to the hospital because they cannot afford the bus fare. A few doctors decided to open their own practices in the villages themselves, but they are afraid they will not have enough money to keep them going. The hospital director commutes from Slatina. Funding and staff shortages meant that some units, such as the intensive care unit, had to close. Yet the hospital is clean, has its own water and energy supply, and there is a sense that some residual discipline from communist times has survived here.

The museum was closed down and some of its spaces were allocated for use by the school. In the dimly lit Culture House, a new generation of dancers, recruited by Nelă, strut about on stage, while the medals won by their predecessors shine on the walls. A little further away, youngsters go clubbing for the evening. In the morning they can be seen working in the fields. The college students still go to work hoeing as in communist times: After all, they study agriculture; for them it is a choice they made themselves. They march along the village road with spades and shovels on their shoulder. Talk to them, and they will complain of the poor-quality water and the dodgy housing blocks. But they would not want a return to the old country ways; they would rather move on to a real town. They are most emphatic in describing the older locals as still being "peasants." Scorniceşti is no town, "'cos you don't have peasants in a town, do you," one of them says, in simple, straightforward words. Yet people are no longer as boorish as their children—who watch *The Bundy Family* and *Ally McBeal* on TV—see them. For instance, the residents of the apartment blocks would no longer want to have to traipse to the backyard to use the toilet. Nor would they want to have to get up in the middle of the night to feed the poultry. In Scorniceşti today, around half of the residents who live in blocks also own land. The average is thirteen hectares per family. The majority own a cow or a plot of land, either themselves or with their parents, so food reserves are secure. This is why they do not complain about their low factory wages. Whoever does not have land, will keep a cow with a neighbor who does. The reclaimed subsistence plots allow families to produce a fair amount of food for their own needs, but few make any money out of it. Yet there is a lot of nostalgia for the old times, as the director of Pulsor says:

> It all peaked between 1985 and '87, production levels were good then, we worked Saturdays and Sundays, and in two shifts, as well as putting in extra hours. After that time, it went down and it keeps going down. ARO, the local car manufacturers, are importing only foreign-made engines, Toyota, Cherokee. We think they could've kept using Romanian-made engines, they were much better. The state does not intervene, taxes are killing us. What they're doing is they're ruining the country's car- and tractor-manufacturing industry and all the related industries servicing it. I've got to tell you, we used to make very good engines. Our engines now operating in Iran on difficult terrains are very good and extremely cost-effective. I'm not saying that the others, abroad, are not making extraordinary engines, 'cos they do, but they cost 80,000 dollars. Ours cost 12,000 and function equally well. But, there you are, there was a will—and it's not surprising—there was a will to practically destroy the national industry in this transition. I don't know what plans they have with commerce, because in order to have commerce the population has to have purchasing power, and they don't. Somebody wanted to make an import market out of Romania, to ruin the country. And it's been a success. I don't know what's going to happen in two or three years' time. Where will these people be working? If there's no industry, no agriculture, what are we going to do? Even tourism will go the same way.

Most towns the size of Scorniceşti have been pushed backward by transition into the most appalling underdevelopment. Occurring after coerced modernization, demodernization has been rapid and frightening. The whole process looks very much like the dismantling of the Scorniceşti greenhouse, an installation worth millions of dollars which fetched less than 5,000 when dismantled and sold for recycling. People have lost their faith in collectivism, whether planned or spontaneous. Individualism and anti-social behavior are on the increase. It is enough to look at the destruction of the irrigation system by peasants who looted the pipes. When asked who was responsible, they say: "people." In a neighboring village, someone took away a piece of the rail track to use it as a ceiling beam, thereby stopping for two days the only daily train service linking the village with the outside world. People can tell the difference between what is theirs and what belongs to other people in Scorniceşti; but they do not distinguish between their own and the public interest. It would appear that the public has been the main casualty of the processes of collectivization and decollectivization.

Notes

[1] The family's name has been changed to protect the source.
[2] Spălățelu, *Scorniceşti. Vatră de istorie românească*, 6.

CHAPTER 7

From the Dependent Peasant to the Citizen-Peasant: The Bases of a Rural Political Culture

The micro-history of these two villages enables us to survey the entire communist and postcommunist period and to unravel the mechanisms used to subject the countryside as planned and implemented by the Communist Party. There is no doubt whatsoever that the destruction of the peasantry as a social class was an intended outcome of the party's policies. The aim was unambiguously to destroy the village as the symbolic site of traditional culture and to replace that culture with a brand-new social construct: a new way of farming the land, a new social organization, and a new tradition, engineered in the laboratories of the party's Propaganda Section. This process is documented not only in the foundational texts of communism, but also in many party documents, some official, others published after the Revolution, as well as in the minutes of many top-echelon party debates about the fate of peasants and villages. The fundamental question concerns the extent to which this project of enforced modernization was successful. What were the long-term effects of social engineering on the social structure of villages and on the peasants' political behavior and their ability to act collectively? The answer emerging from our analysis so far is inconclusive. To provide a more comprehensive answer to that question, I will now cast one last look at Nucşoara and Scorniceşti, and also draw on a nationwide statistical analysis in order to identify the effects of fifty years of social engineering on the peasantry as a whole.

The social structure of the village world during the communist and postcommunist periods

In the nineteenth century Nucşoara was a region with a sparse population in the vicinity of an extensive monastery-owned estate at Curtea de Argeş. Prince Alexandru Ioan Cuza's secularizing reforms triggered an entirely new social dynamic, creating the first generation of peasant landowners, followed by a second generatio n at the end of World War I, a generation of those who did not have land and leased from those who did. The actions of Iancu Arnăuțoiu at Nucsoara and, throughout Romania, the cancellation of farming debts at the time of the Great Crisis were instrumental in creating a class of middle landholder peasants.

The village of Scorniceşti was typical of the plains. Properties there were extensive, but labor productivity was low, and so land was farmed out to peasants who worked both on their own plots and those of the boyars, on a contract basis. After 1989, land restitution consolidated the processes of property-building in Nucşoara, and signaled the start of a similar process in Scorniceşti. Most peasants today are landowners. In Nucşoara, a nuclear family—i.e., parents and dependent children—own between six and eight acres. In Scorniceşti an extended family—also comprising members associated with the nucleus by marriage—has an average of thirteen hectares. Because prior to 1998 it was impossible to sell land legally, many families decided to combine their plots and work them communally for greater efficiency. The differences in property size do not generate significant differences in social status, and as a general rule there is no correlation between status and wealth. One exception to this pattern are the Rroma. At the bottom of the social ladder, the Rroma are largely landless, and unhappy at being treated as second-class citizens by the council and the police. "We don't like to be called 'Rroma'. But the council practises racial hatred when they take no notice of us. At school, our children are getting a good education. When I go to see Mrs Şerban, she sends me away, the same for all of us here. The police are the same. But the council is really bad." (Rroma artisans, Slatina, Nucşoara) As they are the poorest in the community and with the largest number of children, the Rroma are also the most vulnerable, because they depend on the social benefits distributed by the council. And it is not difficult to guess exactly how this distribution takes place. Some mayors believe that the recipients must do some socially useful work in exchange for benefits, and by this they usually mean mowing the hay on the mayor's or deputy mayor's land. As far as the rest of the villagers are concerned, their social homogeneity can be explained by the relatively small differences in property size. Nobody in either village owns more than fifty hectares of land.

So who are the "peasants"? Is everybody who lives in the countryside a peasant? Or just landowners who farm their own land, even though they might live in the city? Or perhaps only rural inhabitants who work their own lands? By 2000, sixty per cent of land in Romania consisted of individual properties (subsistence plots), sixteen per cent belonged to state farms, twenty per cent was the property of private associations, and four per cent the property of commercial companies. Two-thirds of the land thus distributed consisted of plots of less than three hectares, and even those were often divided into non-contiguous plots; the other third belonged to peasants who had resettled in towns. Sixty per cent of land was owned by people over sixty years of age, i.e. individuals with a life expectancy of less than five years given the current standard of living in Romania's countryside. Only nine per cent of those endowed with land were under forty, and therefore physically fit enough to work on their land. Romania's rural population has aged. About fifty per cent of all rural residents had retired by 2000, due either to age or ill-health; the national average was twenty-eight per cent.[1]

Although most peasants are landholders, only a few are self-declared "farmers," i.e. owners producing for the commercial sector. Most plots are farmed for subsistence purposes, with the produce being sold at the market, or swapped for foodstuffs. This takes us closer to a definition of the "peasantry." We could say that peasants are residents of villages, communes, or small towns (or the suburbs of larger cities) whose households survive by working individual plots or gardens and who, consequently, have a peasant lifestyle based on economic self-sufficiency.

Table 3. Rural/urban comparative social indicators

	Urban—Mean (standard deviation)	Rural—Mean (standard deviation)	Population total—Mean (standard deviation)
Age	44 (16.01)	48 (17.77)	46 (17.02)
Education (years)	4 (1.41)	3 (1.31)	4 (1.50)
Income per person	40 euros	21 euros	30 euros
Income per household	65 euros	42 euros	54 euros

Source: CURS 1[2]

The overwhelming majority of rural residents depend on the land for their survival, irrespective of whether they have other employment apart from farming. Even the village intellectuals would find it impossible to make ends meet without their individual plots and gardens. Rural residents are usually cash-poor, but they have been further impoverished in the transition period by annual inflation rates which varied from 20 to 150 per cent or even higher. Pensions and wages have been constantly linked to—but remained just below—inflation, which has led to a considerable decrease in purchasing power, especially in rural areas (see Table 3).

Rural households produce basic foodstuffs, such as eggs, meat, vegetables, milk, alcoholic brews which the peasants could otherwise not afford to buy. All those who depend for survival on what they produce from their own land can be considered "peasants," and they include the elderly, owners of plots, members of associations, and intellectuals, who do not buy food and who produce all they need on their own land. In all these cases, the most important function of the land is to produce food for the peasant family. Any excess land beyond the individual plot and garden contributes less to the economy of the peasant household. And although the family associations in Constantineşti and Teiuş, as well as the larger individual plots in Nucşoara, produce for commercial, as well as for subsistence, purposes, they cannot yet be considered farms. They have little equipment, and their small revenue is almost totally absorbed by the future production cycle. By 2000, the number of individuals on leased-out land was still very limited, to some extent because the relevant legislation did not come into force until late 1994 (see Table 4). The performance of the associations, as well as of peasant tenants, varies widely. Some associations, such as Comagri, produce almost nothing; others, such as Furnica, yield 300 kilograms per hectare. Very efficient tenants are able to supply owners with 600 kilograms per hectare.[3] A common feature of what is delivered to the landowners consists of animal fodder and is not for sale. Many such arrangements do not involve cash transactions at all, and grain is exchanged in the market for seeds or fertilizer. Agricultural production as such comes chiefly from larger farms and from state farms which were still in the process of privatization in 2002. In recent years, various types of subsidy have been offered to large farms, the so-called "agricultural exploitations" of over 110 hectares. In the summer of 2002, for high-productivity grain crops with a yield of 2,000 kilograms per hectare the investment was 250 euros per hectare, including the cost of imported fertilizer. The state attempted to set a purchase price of 3,000 lei (equivalent to 60 euro cents) per kilogram of grain, but wholesalers would offer no more than 2,000–2,500 lei. After a decade or so of state monopoly, grain started increasingly to be traded by the private sector or used for exchange in kind at the market. This, however, did not lead to an increase in its price. On the contrary. The highly reduced price of grain relative to investment means that the peasant economy is still an underproductive system, and only a major increase in productivity can change things. Smaller plots are entirely non-viable. "If someone leases only one acre out to me, it's his own risk; he can't expect more than three sackfuls of produce per acre," says one farmer from Bolintin Vale, a village in the plains. Vegetable, fruit, and vine production seem more profitable than wheat cultivation, but peasants have no capital and are conservative, so they stick to wheat.

Table 4. Uses of landed property

	%
Worked by the owners	74.2
Worked by owners' kin	5.9
Leased out	4.0
Taken over by the association	2.5
Not worked	2.7
Worked by the owner within the association	3.1
Other	7.6
Total	100.0

Source: CURS 1

Productive agriculture is an aspiration for farmers-in-the-making, the majority of whom are agronomists by training. Most peasants are happy just to produce enough to keep their livestock, but even this is no simple affair, although they entrusted the management of the association to the same people who formerly ran the state cooperatives, "because they had experience in the area and know about land." Those who prefer to work their own land do so largely with traditional equipment, the horse and the spade, because, they say, the spade "is good, otherwise if you mechanize you must use herbicide and it's expensive, 30 dollars per acre." The peasants whose chief investment is labor and who use primitive farming tools are in the majority (see Table 4). They are, in fact, the real "peasants," those who have survived as a class, and form a group with shared economic interests. Their contribution to the Romanian economy is, however, nil. The peasantry is merely subsisting, nothing more.

Nowadays, the rural population tends to assess the country's economic situation as being rather good, unlike the urbanites, who have been plunged into the depths of a dark pessimism by a decade of constant decline in industrial output. Yet, even today, rural residents earn only sixty per cent of what their urban counterparts earn (see Table 3). Incomes are more evenly spread in rural areas, and the population is more advanced in age and less educated. Ninety-six per cent of rural residents live in traditional homes, compared with only forty-six per cent of those who have homes in towns of under 30,000 inhabitants. Owning a traditional house, with no running water or central heating, proved an advantage when subsidies for domestic energy consumption were cut during the transition and energy bills spiraled in the towns. Many rural households still do not have a telephone or a washing machine. Private heating arrangements involve having to secure a permit and having sufficient money to purchase firewood; many people are therefore dependent on those who control access to such resources.

Urban–rural inequalities in lifestyle are a legacy of history, and they survived despite communist slogans. Around 1965, peasant income was about half the national average. From 1964, when Nicolae Ceauşescu took power, industrialization under the communists soared at an unprecedented rate. This produced additional incentives for young, active, enterprising people to leave the villages. They settled in towns in great numbers in the 1960s, 1970s, and 1980s. Only in 1996 was migration from and to towns more or less in balance. This happened as migration to the cities declined as a result of ageing among the rural population. It declined from 42.5 per cent in 1973 before peaking spectacularly to 69.8 per cent in 1990, after the collapse of communism, and dropping again to 24.7 per cent in 1996 (see Table 5). Reverse migration, previously insignificant, rose as cities were hit by the economic recession of the early 1990s, and urban–rural migration reached 30.7 per cent in 1999.

Table 5. Urban/rural migration 1973–2000

	1973	1989	1990	1991	1992	1993	1994	1995	1996	1997	1998	1999	2000
U–U	21.9	19.2	18.2	20.2	24.3	25.5	25.6	26.1	27.4	25.0	26.0	26.5	31.9
U–R	10.8	6.5	3.5	10.1	13.7	14.6	18.4	20.8	23.4	26.8	28.4	30.7	27.8
R–U	42.5	55.4	69.8	50.3	39.2	35.0	30.5	25.1	24.7	22.6	22.0	21.0	20.4
R–R	24.8	18.9	8.5	19.4	22.8	25.0	25.5	28.0	24.5	25.6	23.6	21.8	19.9

Key: U=urban, R=rural
Source: CNS

Table 6. Occupational structure of the active rural population (%)

Occupational status	Entrepreneur	Individual	Employee farmer	Association member	Housewives	Other
Active population total	0.1	29.3	55.6	5.2	4.3	5.5

Source: CNS 2000

The percentage of people employed in agriculture declined steadily in the communist period and was reported to be under thirty by 1981. However, by 2000 it was about thirty-five per cent (see Table 6). The percentage of people living in villages, most commonly as families in mixed employment—in which men work in the nearby town, while women work in the garden and on the arable plot—was far higher (forty-seven per cent, see Table 7). In the communist period, as men left the villages to take up employment in the new industries, even if only temporarily or as commuters, women increasingly ended up working in agriculture. Although they represented only fourteen per cent of the national workforce in 1979, women accounted for sixty-three per cent of those employed in agriculture.

Table 7. Urban/rural residence in 2000

Size of towns and villages	%
Town with over 200,000 inhabitants	20.6
Town with 100,000–200,000 inhabitants	8.3
Town with 30,000–100,000 inhabitants	14.3
Town with under 30,000 inhabitants	9.4
Village, commune	47.3
Total	100.0

Source: CURS 1

The transition has strongly affected the social structure bequeathed by communist social engineering. The long-term impact of modernization could be assessed only once the distortions induced by the planned economy had disappeared starting in 2000. One such distortion was the hundred per cent employment rate. At the time, men from rural areas could get jobs as security guards for state enterprises or other such sinecures, but all this disappeared with the total collapse of the state sector. Commuters from rural areas were among the first to lose their jobs in the early 1990s. The lucky ones took advantage of a special scheme offered by the government and applied for early retirement, returning to their villages with pensions that were much higher than what the cooperatives would have offered them. The majority of cooperative members, as well as skilled workers from urban areas—the most typical products of communist social engineering—retired on minimum pensions ten years after the fall of the regime. Half of those who were cooperative members in 1989 had retired within the next ten years, and a quarter of them owned individual plots. Others joined the agricultural associations which succeeded the cooperatives as new forms of voluntary joint ownership.

Table 8. Changes in occupational status during the transition decade (general population)

Occupation	1989	2000
None	1.6	4.9
Housewife	5.6	8.2
Family member	0.4	1.0
Student	17.2	3.9
Unemployed	0.4	6.4
Pensioner	10.1	28.4
Farmer	1.9	5.6
CAP (cooperative) member	6.7	0.4
Agricultural labourer	1.2	0.4
Unskilled worker	4.9	3.2
Skilled worker	31.3	17.6
Shop assistant	7.1	6.7
Self-employed	0.4	1.6
Technician	3.4	2.1
Graduate employee	7.1	7.3
Manager	0.7	0.7
Patron	0.1	1.6
Total	100.0	100.0

Source: CURS 1

Status and upward mobility

There is little social stratification in the rural world. The little that exists is largely a residual legacy of the communist period. The only groups who can claim a distinct status are the agronomists, or agricultural specialists, in lowland areas and the foresters in the mountains. Both groups have considerable power as experts, not to mention other, equally important, powers they have in relation to access to resources and the control of key roles in the agricultural associations.

Who precisely are these groups—the agronomists and foresters—which share a lifestyle and many interests with the peasants, yet enjoy far greater influence and, more importantly, a monopoly on access to major resources? The definition is almost self-evident: The foresters and the agronomists are corporate groups, or status groups. Peasant society in general values status highly. After the social leveling which occurred under communism, distinctions amidst this society are now made on the basis of the groups' closeness to power. The term "status group," as originally defined by Weber,[4] referred to a group of people whose prestige derives from a specific monopoly which they managed to acquire. This community has a shared lifestyle and moral code, as well as its own social philosophy linked to the monopoly they have of specific resources. The birth of a status group, Weber noted, is grounded in usurpation strategies which involve sustained group efforts to acquire access to ever higher status. Status groups have a communal organization, are very exclusive, and protect themselves against contact with strangers. The suggestion that the social organization of Romania was based on closed access and privileged groups was first posited by Ken Jowitt,[5] and later taken up by Mungiu-Pippidi.[6] While Jowitt claimed that this type of closed society was a structural characteristic of Romanian peasant society, I suggested that it was the outcome of social intervention under communism. The annihilation of social stratification by the communists and the positioning of people solely in relation to the sources of power have encouraged this type of social organization, which survived communism and developed further during the transition period. Nucşoara's predatory elite, which included many of the commune's intellectuals drawn from various professional backgrounds, is such a status group, and controls access to all public services.

Entrepreneurs form a special category in the countryside. They speak for the future. Many are descendants of peasants and have returned from the towns. However, rural businesses in building, manufacturing, and trade are run by a category of people with a peasant background. In many cases, individuals in this category spent a transitional period in towns before returning to the village. Domneşti, for instance, the larger commune near Scorniceşti, is full of such people—ranging from forestry engineers to old age pensioners—who returned from Piteşti to their parents' villages once they retired. In Scorniceşti, a great part of the entrepreneurial elite comprises people who spent time in cities, usually for their studies but often after graduation too. They are the only ones who earn cash, employ other people, and create economic growth. In theory, entrepreneurs should be an inclusive category, operating openly in the market economy. This is true of this particular group, but only to a certain extent. Their new lifestyle, which they acquired quickly, and their resources show that in their case there is an overlap between class and status group. Entrepreneurs in Scorniceşti enjoy the hunt and other shared leisure pursuits, and they even have their own club. The people who meet there, the chiefs of police with a monopoly on guns, the local bankers who control access to loans, and the political leaders, form a category which increasingly resembles a status group. Membership of this club, which controls all the major local resources, is not drawn exclusively from one social class. The bankers and policemen come from state institutions and derive their power from the illegitimate control—a de facto privatization, actually—of access to these public resources. The state banks support many businesses by lending money, and the main status group in Scorniceşti gravitates around these loans. But what gives hope to the town is the smaller alternative group of businesses people, those who manage without state bank loans, like Marlene and her factory. If their success is sustainable, it will advance the open society of Scorniceşti as a whole, as they are the real promoters of economic growth.

The two villages being studied here illustrate two different, yet equally significant, trends. Nucşoara, the picturesque village of the mountainous region, has a demographic structure which is typical of the Romanian countryside today: The population is ageing and the young have either left or are leaving. The population declined from 2,000 in 1991 to 1,800 in 2002. There are few weddings and even fewer births. At school, classes keep shrinking and teachers have fewer children to teach. Vacant homes fail to find buyers. Nucşoara is going to disappear gradually; so too are the larger communes in the mountains. In the post-transition period, modernization continues apace: in Domneşti the handsome nineteenth-century village pub was demolished and replaced with bars and terraces with plastic chairs.

The coal mines have closed, and all the men are now unemployed: mowing the hay is their only lucrative occupation. People from the city have started to build in the area, but these are, of course, holiday homes. One way or another, the real peasants will disappear.

In contrast, Scorniceşti, the unappealing village on the plains, full of dust and demolition sites, and located in an even more rural county, appears to be prospering. There are problems with the infrastructure, but although serious they do not threaten the town's future. On the contrary, there are hopes that enough resources will accumulate to encourage growth. There is little unemployment, and the textile industry will be sustainable if some politician does not one day decide out of the blue to scrap it. It is not doing well enough to make people prosperous, but sufficiently well to bring in necessary cash—in parallel to the locally produced food—and to create in Scorniceşti a small market for non-agricultural produce. The only way in which individuals and the town as a whole will survive is if both agricultural and industrial activities can be sustained, until a stronger industry emerges and brings in better revenue. Already, about half of those employed in local industries no longer have land and are, therefore, no longer peasants. The younger generation still live off the produce of their parents' gardens and livestock, but they are losing the skills and inclination to keep a cow on their own. Lifestyles in Scorniceşti are not yet urbanized, but they will be.

The overall levels of poverty drastically reduce the peasants' options. They have little money to buy clothes, and they can scarcely afford to buy newspapers or books. Television is an exception, and the most influential means of urban culture. Yet television programs are perceived differently in villages and towns. The villagers are not keen on political talk shows, which they consider too confrontational, and they generally avoid politics. Soaps and television series, whether set in Spain, Brazil, or Trollope's England, are regarded as good family spectacles, but they are watched in total incomprehension of the context. I asked one of my local respondents for an update on the latest episode of a British period drama, and was told that "the Duke has troubles with his children." Television has an influence on the young, and it often teaches them about things they cannot afford. In Scorniceşti, children have never travelled further than the neighboring town, Slatina, yet the stock reference in discussions among adolescents is to *Beverly Hills 2001*.

It doesn't matter whether villages are moving towards urbanization or extinction, one thing is common to both these tendencies: the decline in the influence of traditional sources of authority. After communism, priests have become mere mortals, and no longer enjoy the respect they enjoyed before the last war. Churches, both in Nucşoara's hamlets and in the fief of the mayor of Scorniceşti, are full of elderly women. Nucşoara's priest sided with those who opposed the arrival of the partisans' cross in the village cemetery. It was surely not a Christian gesture, yet it was not done maliciously, but out of an entrenched conviction that there is such a thing as an "official" truth, or at least that the only truth is the one upheld by a majority. Unsurprisingly, nobody sees him as an anointed guardian of good and evil, but merely as a supplier of essential social services, such as funerals, although even those he has to provide free of charge now.

God is no longer a spiritual and moral reference. Fifty years of atheism, and especially repeated exposure to experiences which have shown that evil can be rewarded or at least victorious on a large scale, have taken their toll: Nowadays only the old former political detainees from Nucşoara ever mention God. When the peasants vote today for the former Communist Party secretaries, it is a quiet profession of atheism. In today's rural Romania, those with a spiritual calling take the vows. Thousands of young peasants have done so in the last ten years, but in the social sphere the Church has got nothing to offer. Nucşoara's priest, a Democratic Party candidate, and one of the many priests in Romania to have participated to political elections, says honestly: "This was chance, you know. I chose this party because they came to me and suggested it." So, even when it comes down to opportunism, the Church will let others come to it rather than take the initiative. The vacuum left in the social arena by a lethargic Eastern Orthodox Church has been filled by the very active Protestant sects. In Scorniceşti they recruited enough adherents to be able to start building a church.

David Kideckel noted with surprise that neither the Church nor the family were able to protect the individual against the compromises demanded by the communist regime.[7] It would be more accurate to say that they were no longer able to do so after the 1950s. The appointment of Patriarch Justinian as head of the Romanian Orthodox Church in 1948 marked the beginning of an accommodation between the Church and the communist state. Not unlike Nucşoara's peasants, who were endowed with other people's lands, the Church was offered incentives and became a partner in the regime's success: It took over the assets of the suppressed Greek-Catholic Church, hundreds of churches for the return of which Greek-Catholics fought bitterly in court after 1989. The Eastern Orthodox priests who joined the resistance did so as a personal choice. Afterwards, out of *esprit de corps*, they recruited one another to join the cause of the partisans. Father Dragomirescu served a long prison sentence simply because the partisans asked for a favor—which he could not even provide. The commune's other priests, Andreescu, Constantinescu, and Drăgoi, were all executed by the communist regime. It is unsurprising, therefore, that their successors should have refrained from giving advice on moral and political conduct to their parishioners. They were content with simply saying a prayer for the well-being of the state's leader, as Orthodox tradition required, even when the leader in question was Nicolae Ceauşescu.

The role of the family was equally ambiguous. In the postcommunist transition period, the rural family has appeared to gain strength chiefly as an economic unit, as family members gradually realized that only by combining their labor could they exploit their land productively. Under communism, however, stories of family solidarity are matched in equal numbers by stories of betrayal or indifference. Marina Chircă is still angry with her sister, Ana, because the latter allowed herself to fall for a Securitate man posing as a suitor and let slip a few names, which later led to the partisans' capture. By the time Ana realized it was a trap, it was too late. Having learned this, Marina left her house the next day without taking anything with her, determined never to return because she was certain the Securitate were about to arrest her. As she approached the village bridge, she came across Ana, and took her along with her. Ana did not question anything, left her Securitate suitor behind, and together they went up into the mountains, where they hid for five years before they were caught. Despite having to share the same attic for five years in the house of a former servant, they were never reconciled.

Family ties sometimes break down under the pressure of dramatic circumstances. After a hesitant start, the Securitate came up with a well-defined strategy, especially in the capture of partisans: It recruited members of their families, who were then subjected to blackmail and torture. These "deep-cover informants" were often credited in Securitate documents with the capture of opponents of the regime. Once the purely coercive phase came to an end, the family was often invoked as an excuse and cover for selfishness and compromise. "Father joined the party for our sakes," said the doctor in Nucşoara. Such self-exonerating statements are shared by the four million former members of the Communist Party and their families, whether peasants or not.

The neo-dependency model

Communism destroyed the traditional sources of authority and created new dependencies, on a different basis. Under the communist regime, there were three main sources for the dependency of the peasant either as a cooperative member or as a commuting employee in a state enterprise. On the one hand, there was an economic dependency, not just on salaries, pensions, and allocations, but also, for cooperative members, on food rations. In the 1980s a rationing system was also introduced in state enterprises, and consequently access to basic foodstuffs such as sugar, oil, and bread became strictly controlled and correlated to party membership and discipline.

The second source of dependency was political, and was more widespread in urban areas than in the countryside. In villages, too, state employees—intellectuals, civil servants, cooperative members, and state farm staff—were subject to political control. Debate was allowed only within the system thus created, as illustrated by the very suggestive recollections of Nicolae Popa from Scorniceşti. Popa, like Constantin Paşol from Nucşoara, was a communist from the start. Both attempted to encourage a debate within the system, by denouncing corruption and suggesting alternatives to the party's absurd policies. Significantly, neither was successful. Political dependency was made complete through a network of political organizations membership of which was compulsory: the Union of Young Communists (UTC), the trade unions, and the Communist Party itself.

The third source of dependency is the most important for understanding why the status groups survived and even flourished during the transition. It is what Waldron has called personal dependency on one's superiors.[8] Far from being impersonal and equitable, communism, like any other tyranny, produced smaller tyrants who lived off the arbitrary distribution of promotion and awards. The overblown personal character of the system, the excessive growth of string-pulling, and old boys' networks were the rule rather than the exception and have had long-term consequences. The patronage networks of Romania today are firmly grounded in its communist past.

The new personal relations of dependency are undeniable, and they replaced fully those based on the traditional authority of prewar times. As a priest's son and a teacher, Arnăuțoiu enjoyed real respect from his fellow-villagers. His power stemmed from his affiliation to these categories of traditional authority, strengthened by his considerable personal charisma. The same can hardly be said of Ungureanu and Neacşu. Their past and present power rests on their proximity to a discretionary center of power, whose agents they have always been. The authority they exercise is not their own, and legitimacy is out of the question: They live on the residue of a coercive power over their fellow-villagers, converted into a milder form of patronage. Authority normally means that "A has legitimate power over B, with the latter's consent." However, under communism, and even today, in the postcommunist era, when peasants do not have an alternative source for the goods they receive from the local authorities, we can hardly talk of free, consensual social relations. In this case, what we find is domination, the sustained exercise thereof, and not authority, which in all its forms is constantly deflected and corrupted. This is the reason I call the postcommunist model of social control described here a neo-dependency model.

There are connections between neo-dependency—with its implicit particularism and the non-modern character of its economic and administrative bureaucracy—and the communist regime's earlier, coercive phase. Captors became patrons, and pure coercion was upgraded to persuasive coercion, but they remained the principals of this game. This second phase of Romanian totalitarianism was not far from what Waldron, based on his ethnographic work, defined as an alternative model to totalitarianism and labeled as "neo-traditionalism." Yet, this phase appears to have been simply a developmental stage of the same old totalitarianism, the main difference being that the blind, cost-intensive coercion of the earlier stage was replaced by a rational contract offered to those who accepted the system both publicly and in private. If this contract is seen as a means by which the regime acquired legitimacy, then we have indeed a regime which differs from the hard-line totalitarianism of the Stalinist era, i.e. a neo-traditional regime. However, the Romanian experience contradicts such an interpretation. Neo-dependency is a subtler way of controlling society as a whole. It is both more comprehensive and more pervasive than coercion. For this reason, a neo-dependency regime needs a greater number of cadres: Every head of section, civil servant, or party secretary becomes virtually an agent for society's mobilization and control. The arbitrariness of power and the particularism are not side-effects, but signals and manifestations of the unlimited power shared by those networked to the central source. There will be hierarchies, naturally, separating the top chiefs from people such as Mrs Şerban, but her power will endure as long as it is discretionary and, as such, each agent of the regime will behave just like her. Abusive attitudes are exterior signs of power, and particularism creates the necessary incentives for not rebelling against an unjust system. It is easier to network than to revolt! In the social arena, energies are not directed towards changing the rules of the game towards a system based on open and non-personal relationships. Instead, efforts concentrate on climbing the ladder towards inclusion in the group with access to the source of all bounty. This mechanism deflects social energies towards opportunism rather than social protest. There will still be those who do not choose this path or will openly defy the system. Against those, the neo-dependency regime will employ surveillance and repression as relentlessly and efficiently as any totalitarian system. Yet, repression will be less visible because it will be directed against fewer individuals, the majority having accepted the regime, a process first analyzed by the Romanian dissident Mihai Botez.[9]

Many of the former political detainees had to resign themselves as well: The regime was there to stay. The man who opposed collectivization in Constantineşti ended up working closely with Vasile Bărbulescu at the cooperative, where he was employed as a driver.

One of the sons of Father Drăgoi became head of the local Forestry Service, because he was better educated than the others and the regime needed someone with the necessary accountancy skills. In the 1960s their choices had been limited: The young ones were barred from higher education, as were those children considered to have come from a kulak background. However, in the 1970s they started to be gradually integrated into the regime. The most spectacular case was that of the partisan Ion Gavrilă, who managed to hide until the 1970s, when he was arrested. Ten years earlier, he would have been put on summary trial and executed. But the Securitate officers who processed his case were a new generation, with better education and subtler persuasive techniques. His principal job-searching ally was a Securitate colonel, who eventually, after many failed applications, managed to get him a job. He was hired by a state farm in Sibiu county. There, the former partisan pulled his weight, showed creativity in responding to his superiors' absurd demands, and became an important cog in the socialist machine.

In rural areas, political power belonged to a narrow status group, which held it on behalf of the communist regime. They managed to do so by taking advantage of one essential aspect which remained unchanged: the peasants' lack of economic autonomy. In the place of cooking oil or sugar rations, these rural elites distributed property title deeds, farming subsidies, and permits for various projects, which in other countries are either unnecessary or are issued automatically and non-discretionarily. In postcommunist Romania, however, such applications are dealt with so slowly and ineptly that applicants need a personal connection to secure a result.[10] This is a self-perpetuating system: The superior status group is the only agency capable of oiling the state's chronically cranky machinery, on which the subsistence peasant economy depends for survival.

In a society such as the one described here, power becomes particularistic and creates relations of dependency and deference, which is the contrary of modernity, where authority is universal and non-personal. The privileged groups subsequently invest in pseudo-authority, that is, they make and unmake subaltern authorities according to the group's interests: "We will elevate so-and-so, but we don't need so-and-so." Mayors are frequently appointed in this way: They are often nobodies with little authority, and people know that they are controlled from backstage by the real power-holders, to whom sole deference is due. Nobody in the two villages discussed here ever suggested for one minute that their mayors are the most influential men in the village. In Scorniceşti's local council, Mr Neacşu commands more respect than the present mayor. As for the mayor of Nucşoara, people are quite happy to tell you that the mayor is "retarded" and was placed there only because he is easy to manipulate.

Once created, pseudo-authority strives to gain legitimacy and recognition on a scale compensating its modest start. Ceauşescu, for instance, demolished the houses of the old Bucharest elite to build a brand-new neighborhood, which included the People's House and the Boulevard of the Victory of Socialism. The entire nation was his playground. Elena Bărbulescu and her husband had only Scorniceşti, but they did what they could: the "Brooklyn Bridge," the 20,000-seater stadium, accommodating more than the village's entire population, were no mean achievements. The post-transition nouveaux riches come from the city, where they made their fortunes, and go to the countryside to build not houses, but palaces. Theirs are palaces in concrete, with neon lights and a remote control for the garage, for they are not sufficiently well educated to have good taste or employ an interior decorator. There are not many of them, but enough to make a pattern. The most famous is the tycoon-turned-convict Ilie Alexandru, who had a detailed, smaller replica of the *Dallas* Southfork Ranch built in the Bărăgan region. Each village seems to have its own Ilie Alexandru. This shows that the village remains the symbolic site for converting wealth into status, the new elite's most prized currency.

These local political power holders, located in the village and interacting directly with the peasants, must be distinguished from the real principals, of which they are the agents. The local authority can be said to sit astride a frontier line looking in both directions. It is there to relay power, which is why when communism was installed it was so important to gain this strategic position and secure a transfer of power from the old power holders such as Nucşoara's Arnăuţoiu family. In order to do this, as elections loomed in the crucial year 1946, the Communist Party brought in the heavy artillery: in Scorniceşti this was Ceauşescu himself. The Communist Party secured by violent means its privileged position at the interface between the village and the outside world. To achieve this, the party created new cadres overnight. People like Gheorghe Şerban were quickly installed in key posts and trained on the job to give them a smattering of legitimacy. After 1990, the same position was disputed by representatives of several parties. But was the competition fair? No, because the key to success remained proximity to state power, which was firmly in the grip of former communists. This explains the considerable ascendancy gained by people such as Ungureanu, the *éminence grise*, the most influential man in Nucşoara. He joined the main postcommunist party on its creation and had a nephew who became deputy for Argeş county. These were clear signals for the villagers that he had the key which provided access to the new power center as well as the old one. Later, that nephew became a prefect, which only enhanced the informal authority already enjoyed by Nucşoara's chief gatekeeper. The retired forester Ungureanu denies that he is the source of power, and argues that the favors he did for people, either himself or through his godfather, the Securitate colonel Ploscaru, endeared him to them.

Ungureanu's wife had been christened—along with half of the village—by Iancu Arnăuţoiu. Later, Ungureanu and Ploscaru became in-laws: The Securitate officer had come to the village to hunt down partisans, but ended up sharing their life as one of them. This maneuver is more than an opportunistic survival strategy, it is the very essence of neo-traditionalism. It involves adapting a form of traditional alliance to an entirely new situation and—because Ungureanu had been recruited because of a blemish in his personal record—it also meant turning coercion into collaboration, as well as into a social and family contract. His position today is a somewhat humbler variant of the position held by Iancu Arnăuţoiu before the war. This is how Arnăuţoiu's daughter sees the situation:

> Father was much loved and listened to in the village. They'd all come to him for advice. When someone was about to marry, they'd come to consult with father. Nobody ever signed a document without seeing father first. He'd make the draft and then they'd go and have that certified. And I can tell you, at one stage, over half the people in the village were my father's godchildren! [...] And father said to them: "If you vote for this party now in power, they'll build a school for you." And they'd all go to vote, my father and the whole village, and then they had their school. The parties kept their word to him, and he

> bargained with them. Because, I can assure you, father's word had some weight in the village. And when father went with the whole village to cast their votes for those guys—and everyone in the country did it—of course the winners built their school. He managed to have schools built this way in Secături, in Sboghiţeşti, and in Slatina. Three schools![11]

Iancu Arnăuţoiu mediated between the village and the central source of power: He brokered loans for the purchase of land, he was everybody's guarantor, and it was he, too, who fixed the votes, cast en bloc according to previously made arrangements. He did not have power, but he exercised a legitimate authority recognized as such. He was with the Peasant Party: Ion Mihalache and his wife had honored this allegiance by attending his daughter's wedding. But in the "votes-for-schools affair" he voted with the liberals, the rival party, because they had a better chance of winning and had promised him the schools. Ion Mihalache famously complained of people's compliance to this system. He identified four central factors which corrupted political participation in villages: One was pure opportunism (you join the party which gives you something in return), the second was "functional" partisanship (the need for political power as a weapon against a rival or enemy), the third was seduction (you go for the most "attractive" of the candidates), and, finally, kinship politics. In his own words: "I am conservative because this is what I think is best, because the Godfather is conservative and I cannot cross him, because that would mean being against him." "Our politicians know our weaknesses full well. They have started a real 'hunt' in the villages, marrying, christening, wooing village leaders only for political purposes. They are not the culprits; the culprits are those who indulge in this political slavery, an undignified position for any reasonable citizen endowed with free will. Without doubt, politics [...] has become a matter of seeing to personal interests at the expense of public interests."[12]

Ion Luca Caragiale wrote similarly about the "historic" (conservative and liberal) parties, which, he contended, were "two large factions, each with their own clients rather than partisans."[13] This system, dating back to the era of limited suffrage, was recreated in Romania in the aftermath of the introduction of political pluralism in 1990. But the country, as we shall discuss further, had only partly been freed from communism at that stage.

The reinvention of politics in the village

Brutal coercion, political as well as economic, which was revived in the 1980s with the new delivery quotas for livestock, suddenly disappeared in 1990. The peasants received their plots back, without any further conditions. This explains why the peasantry as a whole voted for Ion Iliescu and why the peasants were slow to realize that the source of all this bounty was the Revolution itself, not its leader, enthroned by the army and the bureaucrats. In the village world, evil, as well as good, is personalized along the lines of the old traditional authority. This is best expressed by Nicolae Popa when he compares Dej to Ceauşescu: "Dej was a kind man, a very kind man. He had our time in the army added onto our employment cards and, himself being uneducated, he introduced that scheme with education at the workplace. Then Ceauşescu came along and took those years off our length of employment. You could say that one God gaveth and another tooketh away."

The children of Scorniceşti had not yet been born when Ceauşescu was alive, but they hero-worship him: Ceauşescu was the Leader with a capital L; of course he had his shortcomings, but he was true leader material. Mr Paşol is highly dismissive of the leadership qualities of anti-communist politicians. "These guys couldn't, even if they would," he says. The only one to come close to Ceauşescu on the leadership vigour scale is Vadim Tudor, the populist radical leader, who was in fact one of the communist poets praising Ceauşescu. In Scorniceşti, he is well liked by the children, by the communist Constantin Paşol, as well as by the teacher Marinică Popa. The latter, although generally critical of collectivization and communism, reminisces fondly about the "leadership skills" which Ceauşescu definitely had, in spite of his "lack of schooling."

In legislative elections, in Scorniceşti and Nucşoara alike, the vast majority voted consistently and without exception for the "governing party," i.e. the party regarded as the successor to the party state. Initially named the National Salvation Front (FSN), it was subsequently renamed the PDSR (Socialist Democratic Party of Romania), and then the PSD (Social Democratic Party). This voting pattern prevailed in most of rural Romania, with the exception of a few highly developed rural areas such as Prahova, which has fewer peasants owing to the fact that the villagers there are employed in the petrol industry and in tourism. At most, when one or two members of the local elite decide to leave their comfort zone and try another party, some anti-communist councilors might get elected, but never more than a third of the total. Miraculously, the liberals managed to win in Scorniceşti one year, but this was a mere blip, and the general elections were again won by the PSD, represented by Mr Neacşu. If you ask people why they voted for him, they tell you: "Well, what else could we do, since he ran!" But the former Communist Party activist is not without merits. The local council in Scorniceşti has several EU programs in place, some initiated when Neacşu was mayor, others under his designated successor, the current incumbent. By contrast, Nucşoara is stagnating. During the electoral campaign, all the candidates explained that there was no money then, nor would there be in the future—something which the villagers already knew—because there were very few salary earners and the council could not collect enough revenue from the pensioners alone.

But how did political freedom start? Like everywhere else in Romania, it started in an atmosphere of revolutionary fervor. In Scorniceşti, the first mayor "was not elected, he elected himself. He simply went to the village hall on the first day of the Bucharest revolution and remained there until 1992. In that year, elections were finally organized, contested by eighteen candidates, and the one with the most buddies won. They had no political platform, just promises." In Nucşoara, the first to be unanimously nominated was the cashier of the cooperative, but he withdrew as soon as he realized what a headache the land-related conflicts were going to be. In any case, he was one of the wealthiest people in the commune and he did not need the job. Then, for about nine years, there was just one mayor. Initially he was with the governing party, but when the latter split he, like Ungureanu, the prefect, joined the party's younger branch. As regards the councilors, they came predominantly from the older post-communist party. The only political changes occurred not through shifts in voting, but through the migration of councilors. In the words of Nucşoara's first postcommunist mayor, the one who stepped aside: "The councilors migrate from one party to another. But the secretary [Mrs Şerban] is still here and she was here before" [i.e. before 1989].

On the rare occasions when something new does emerge, it ends up being controlled by the old guard, because the predatory elite have ample means to hijack any project. A young female teacher from Scorniceşti—not young enough to have been spared agricultural "duty" with her pupils under Ceauşescu—took very seriously the words of a leader who left the governing party to do something new:

> You know, when [...] said he was going to leave the PDSR and create his own party, with younger people. But there was no-one to help build the ApR [Alliance for Romania] in Scorniceşti. And he'd keep calling me to Slatina, to talk about things. "Wouldn't you want to do this for the ApR? You know the people, you can convince them, and we're going to pay you." And I said "As far as I'm concerned, I just want presents for the children, for Children's Day, and for Christmas." That was all. And so, I went out to talk to people, I recruited members, I recruited [...] the most members. Younger people. And then once I've made the party—very nicely, we had a party at the community center, and I washed the dishes and bought the food and cleaned up—then the deputy mayor comes and takes it all from me. And then this party leader actually increased my membership fee and insisted I attend their meetings in Slatina. So I said what the hell, if you're no longer happy with my work, you can have your bloody party back. And it used to be such a good party.

Electoral campaigns are quiet affairs, but there is plenty to drink and a raft of promises. The Rroma are carted off to the polling stations in the candidates' cars to make sure they vote for the right people. Nucşoara's former mayor says that the local branch of the PDSR blackmailed the Rroma into thinking they would no longer be allowed to collect firewood and timber in the forest if they did not vote PDSR. The party state is represented locally by its chairman, Mr Cornel, a forester, naturally. The Rroma are quick to defend themselves: "Well, they made us promises, so we voted for them. We asked for three neon lamps, and we didn't even get that. We would have voted for them anyway. Well, yes, they come and turn our heads with a bottle of brandy and promise they're doing this and doing that. They come and take us [to vote, *author's note*] in the car. I was recovering after an operation and they still came to take me. Mr Berja [the current PSD mayor, *author's note*] came with a light lorry to take me. But the candidates from the other parties were here too." In Scorniceşti, too, the mayor distributed serious amounts of booze during the campaign, and made promises to the local grandees. So it goes. However, the much-maligned Rroma are fully aware ideologically: "We voted with the left 'cos that's what poor people do; it's a poverty vote, and we go with the state, 'cos in the end it's the state that provides for us."

We go with the state, we vote for the state, because only the state has got the power to help us. Although by the time of this conversation anti-communists had succeeded in being returned to office for a four-year term (before being replaced again by former communists), many people still believed in the omnipotence of the state and nobody was in any doubt whatsoever about the predatory elite's ability to hold the state captive. There are two main reasons for this. One is that the anti-communists managed to win only as part of a broad coalition, which was insufficiently successful to be able to claim identification with the state. The second, and very powerful, reason is that locally, in the countryside, state is identified with the status group, i.e. with the predatory elite. This reality is difficult to ignore in Nucşoara, where the PDSR has its headquarters right in the village hall, whereas the Peasant Party has no headquarters at all, and failed to secure a base even while it was in power. "People didn't join, because they were afraid," says Vică Berevoianu, the local peasant leader. "I joined, and so did the mayor's father [a former political detainee, *author's note*], as this was our party before the war. But when the son became mayor, he voted for the PDSR. If you can't beat 'em, join 'em." Iancu Drăgoi, a former partisan, agrees: "People are afraid. When our men were being hunted down, the Securitate would come and load up truckfuls of people, for no other reason than perhaps having run into them in the mountains."

Though coercion might no longer be applied, fear is certainly still present. But is fear the only explanation? Fear cannot entirely explain the retrospective admiration for Ceauşescu, nor can it entirely explain voting behavior. It is easy to see that people are reluctant to side openly with the opposition. The ballot is secret, yet, as the Rroma say, the village voted en masse "for the state." While being typical of a certain way of thinking, this expression is bizarre in that context, because in the elections referred to in the interview the PDSR was returned to power after its only period in opposition (where it had been sent by urban voters in 1996). When interviewed, the peasants say that the governments prior to 2000—i.e. the anti-communists—were better for them. The vouchers handed out by those governments, which the peasants could use to buy fertilizer and pay for machines, proved to be of better practical use than the cash subsidies offered by the PDSR, money which did not always reach the intended beneficiaries. Yet, in spite of all this, the peasants continued to vote for "the state." Here is the explanation offered by a former Communist Party secretary from Scorniceşti:

> Well, yes, under Ceauşescu I used to be local branch secretary. Before, politics meant working for the party, and I joined out of conviction, honestly, out of conviction. And I was, how shall I put it, I was, well, disappointed in a way. I didn't expect all that happened at the time. Now, however, from a different perspective and with all the information that came out, I can say that we were, how shall I say, the product of the society in which we lived. I started talking to people here, and we realized that had we continued under communism we would have gone bust. But, I can tell you honestly now, having done a degree in political studies I'm less of a communist than some people here. The factory workers here, the peasants, now, they really *are* communists. In their hearts.

Are peasants really more "communist" than city-dwellers? And is this the most enduring effect of communist domination? The concluding section of this chapter looks at the evidence.

Rural vs. urban: political cultures compared

The peasants are more skeptical than the urbanites when it comes to the possibility that they might actually have an influence on the politics of their country, region, or village. And yet, they have more faith in the performance of governing bodies than the urbanites do. The differences are statistically significant, and confirm the profile of the city as more active and demanding and of the village as more obedient and passive. Although economically poorer, the village has richer resources of social confidence than the city: The peasants are significantly more inclined to credit all public institutions for their performance (Table 9). With respect to interpersonal trust, the villagers are significantly more reserved with strangers than the city-dwellers are. They are, however, less nationalistic with respect to territories, given that the village is more parochial and identifies with the nation to a lesser extent than the city (Table 10).[14] Otherwise, the urbanites and the villagers are equally sceptical with respect to their fellow humans and believe equally that family and acquaintances are more trustworthy than strangers.

Table 9. The map of social confidence in the village, commune, and city

	City	Commune	Village
Confidence in political parties	13	24	21
Confidence in the local council	49	60	60
Confidence in the army	71	79	77
Confidence in the Church	85	94	93
Confidence in the president	44	54	51
Confidence in the government	38	47	52
Confidence in the justice system	28	39	41

Source: BOP Poll, Metromedia, 2001

Table 10. Political beliefs and values: a village/city comparison

Beliefs and values	Urban	Rural
Communism was a good idea but was inadequately put into practice	54	61
People have to obey their political leaders even when they are wrong	38	46
There are Romanian territories within the boundaries of other states	71	63
Let's get rid of elections and Parliament and place a real leader at the helm	24	36
Most people behave well only when governed with an iron fist	51	57
It is always the same people who benefit from regime changes, whereas decent people continue to have a hard life	82	82
Romania is a beautiful, rich country, but has many enemies who prevent it from prospering	56	58
The state has a duty to provide jobs and a good living standard for everyone	76	85
Equality is preferable to freedom	38	50
Romanians are more trustworthy than foreigners (interpersonal trust 1)	34	46
You can trust only your own family (interpersonal trust 2)	73	76
You can trust only people you know well (interpersonal trust 3)	61	66

Sources: CURS polls, March 2000, CURS 1, Eurobarometer CURS, October 2001

Differences in levels of trust are confirmed by in-depth analysis (Table 11). In a bivariate regression analysis, interpersonal is correlated with rural residency. But if we control for education, income, and age, the association between lack of trust and rural residency no longer applies. In other words, if two individuals have similar levels of income, age, and education, the fact that one lives in a village and the other in a city is not a predictor of a gap in interpersonal trust. Urbanites and villagers are equally distrustful in Romania. If there is such a thing as a specifically "peasant" psychology of distrust, then it can be said to apply to the entire Romanian nation, not just to peasants. The same could be said of envy and social frustration: The variable "It is always the same people who benefit from regime changes, whereas decent people continue to have a hard life" reflects the opinion of vast sections of the population in villages and cities alike.

Table 11. Analysis of interpersonal trust

	Non-standardized coefficients regression (standard error)
Income	0.000 (0.000)
Age	-0.063 (0.028) *
Education	0.024 (0.011) *
Commune/ village	-0.092 (0.068)

Dependent variable: interpersonal trust
Significance levels: *p<0.05, **p<0.01, ***p<0.001
(data from CURS 1 poll)

Two of the supposedly chief features of the peasantry, distrust and envy, are equally observable in urbanites. Predictably, the psychology of the limited good became a cross-national psychology after communism. The Romanian historian Nicolae Iorga was among the first to deplore the absence of solidarity and the lack of self-organizing skills among peasants, features which he attributed to the parochialism of poorer villages and which, he thought, were responsible for the failure of peasant revolts: "Smaller villages [...], scattered across wide distances and almost without connection between them, are like self-sufficient, organic, genealogical cells. This separatist spirit has dominated them from the very beginning, and it is this that makes the peasant individualistic, unable to reach out to his peers and join them in a common endeavor. Romania had peasants, but not peasantry as a social class."[15] Nowadays, such characteristics have spread across postcommunist society. Lack of capacity for collective action translates into very low membership figures for associative life both in the countryside and the city. Social capital indicators, such as trust and associative life, strongly correlate not with rural residency, but with the region's development level, using a regional infrastructure as proxy.

Significant differences between rural and urban can be found in the case of two other important elements of "rural psychology." Peasants are significantly more authoritarian and collectivist than urbanites. They believe that equality of income among people is more important than freedom. They also believe to a greater extent than urbanites that in theory, although not in practice, communism was a good idea. They also believe more than city-dwellers that the state should manage the economy (peasants are more "statist"), but only by a small margin. When we regress either authoritarianism or collectivism on social status controls, it becomes obvious that differences between the age groups, income, and education of urban and rural populations, and not residence in itself, explain their different attitudes. Most differences in Tables 9 and 10 arise from the fact that the population of the countryside is older, less educated, and poorer than its urban counterpart and, therefore, variations in trust and confidence levels are not structural, but related to social context. In some cases, after controlling for income, education, and age we find that the peasants are significantly less statist than the urbanites. In complex analyses which attempt to explain the choice of individualism over collectivism, rural residency does not count significantly, but the development level of the region does.[16] Controlling for income, education, and age, peasants are economically more optimistic and also less nationalistic. Their comparative economic optimism can be explained by the greater autonomy they have vis-à-vis urbanites in terms of producing their own food. Only one genuine rural/urban difference survives, predictably along the lines of a rural psychology: rural residency determines obedience, even when controlling for income, education, and age. The peasants believe that leaders must be obeyed even when they are wrong.

Table 12. Rural residence as a determinant of political attitudes

Dependent variables	Rural	Education	Income	Age
Communism a good idea	Not significant	–	–	+
Statism	-0.05**	–	–	Not significant
Obedience	0.10***	Secondary school and college, *for*; higher education, *against*	Not significant	+
Territorial nationalism	-0.42***	+	+	Not significant
Economic optimism	0.41***	Not significant	+	–

Significance levels: *p<0.05, **p<0.01, ***p<0.001
Source: Author's own analysis based on data from the CURS 1 poll.
The values in the second column are standardized regression coefficients.

Differences in terms of political literacy are perhaps the most striking. Urban residents are rather ill-informed on and uninterested in ideology, but rural residents are even worse in that respect. It is hard to have meaningful politics when only five per cent of respondents think it is important that a candidate is left wing or right wing. The percentage of those who cannot tell the difference between the right and the left explains why this variable does not explain political preference in national voting models. This percentage represents over three-fifths of the rural electorate, if one adds those who said that they know the difference between right and left but then give the wrong answer when tested.

There is a strong urban/rural difference in voting behavior. In both 2000 and 1996, 45 per cent of the rural vote went to the main postcommunist party, the PDSR, perceived as the successor of the party-state, against 32 per cent of the urban vote. Earlier, in the 1992 and the 1990 elections, the percentage of peasants voting for the postcommunists had been even higher, but that was later affected by splits in the new "party state" in 1992 and in 1997 which created much confusion. Rural residence was and remained a strong predictor of vot-

Table 13. Urban/rural political cognizance: a comparison[17]

Characteristics of political cognizance	Urban	Rural
Follows the election campaign in the papers daily	23	14
Followed the election campaign for more than one hour on TV the previous day (in 2000)	32	16
Vote influenced by knowledge of whether a candidate is of the right or the left	9	5
Does not know whether it is the right or the left which supports narrowing income disparities	41	48
Does not know whether it is the right or the left which supports an economy based on private property	39	47

Source: CURS 2 poll

ing in favor of the postcommunist president, Ion Iliescu, even in multivariate models,[18] after controlling for income, education, and age. The strongest correlation between rural residency and a consistently procommunist stance was found, predictably, in the poorer areas in the hills and the plains, which are predominantly agrarian and where there are only a few, recently established towns.

The contrasting findings about peasant passivity and obedience to authority on the one hand and peasant contention and passive resistance on the other have formed the subject of a vast literature, often with contradictory conclusions. In the absence of political freedom and outside support, the peasant would appear to be entirely dependent on those who have absolute control over the state and power resources. Under collectivization, the landless peasant was an example of this dependency at its worst. Conversely, in communist countries, the peasant who had land and weapon and was consequently autonomous defended himself against the predatory communist state, which attempted to reduce him to servitude. The peasant resistance to communism in Romania, animated by the old elite in places such as Nucşoara, or led by peasants themselves—for instance by the Gavrilă group in the Făgăraş mountains—depended on the promise of external support for regime change. It collapsed soon after the Hungarian Revolution of 1956 and had made it clear that the West had abandoned Eastern Europe to the Soviet Union. In the circumstances, regime change was recognized as unattainable.

Even in regimes which guarantee political freedom, one can hardly speak of a peasant political culture. Even when he is enfranchised and the ballot is secret, the peasant will rarely vote against "those in power," although some peasants in impoverished Southern Italy have been known to vote for the Communist Party. Usually, in post-traditional societies, rural voting behavior is that described by Huntington: "The role of the city is constant: it is the permanent source of opposition. The role of the countryside is variable: It is either the source of stability or the source of revolution. [...] He who controls the countryside controls the country."[19] Rural voting is brokered by the local elite. At the time of Iancu Arnăuțoiu, rural voting was a collective rather than an individual affair, and remained so under the incomplete freedom of the postcommunist transition. It has been and continues to be negotiated between the local authority, which controls access to local resources, and the national authority, in a two-tiered model of state capture. Was this model building on the institutional legacy of the interwar model, and did communism become a neo-dependency system under the corrupting influence of a peculiar Romanian mentality and of residual kinship politics? Or was the model built anew? Given the massive and violent impact of communism, the first variant looks unlikely. If we want to understand the parallels between interwar and postcommunist political behavior, the cause must be sought elsewhere. Our survey so far has produced enough evidence to show that we are faced with a reconstruction of the institutional framework of political subjection and particularism. To a large extent this is due to the fact that some of the social conditions which prevailed in the interwar period remained unchallenged in the postcommunist period: There has been no attempt over the last decade to pursue agrarian and rural policies aimed consistently at removing those conditions. A review of the main factors which have caused the postcommunist transition to replicate the interwar political model would look as follows:

1. The reintroduction of subsistence plots, which led to land fragmentation
2. The continued existence of an agrarian overpopulation with no employment opportunities in industry or the service sector
3. The low productivity of individual plots, over fifty per cent of

which are still farmed in traditional ways (using horses for example)
4. An overproduction of poor-quality grain, massively undersold with respect to investment per hectare
5. A dependency on state subsidies in the form of cash or voucher handouts
6. Local gatekeepers having discretion in the distribution of state subsidies and who trade them for votes
7. Political illiteracy and political apathy which, after fifty years of single-party politics, became the equivalent of the interwar illiteracy.

This list suggests how informal arrangements of power in pre-communist Romania reproduced themselves in the postcommunist era: The land policies of the 1990s, on top of the chronic underdevelopment and lack of education, recreated the set of factors which made competitive politics of little meaning in the countryside when tried before the war. The postcommunist transition seems therefore to have regressed the village's institutions to the premodern agrarian and political realities which prevailed prior to 1918. Sadly, that system, denounced by both Mihalache and Caragiale at the time, is even less fitting in the twenty-first century in a country which has now become a member of the European Union. The outcomes alone are more different, but the difference does not favor the present. Past political patronage led to the building of schools and the development of the village. During transition, resources were definitely not used in the public interest, but towards purely private objectives: Patrons are more decidedly predatory spoilers and people victims rather than citizens.

Notes

[1] Data from the National Statistics Institute, available at www.ins.ro.

[2] The samples used are described in Appendix 1.

[3] In the Bărăgan region. Interview conducted in the village of Ciorogîrla, Bolintin Vale, Ilfov county.

[4] Weber, *On Charisma and Institution Building*, 177–80.

[5] Jowitt, *Social Change.*

[6] Mungiu-Pippidi, *Politica după comunism*, and idem, *Secera si buldozerul.*

[7] Kideckel, *The Solitude of Collectivism.*

[8] Waldron, *Communist Neo-Traditionalism*, 21.

[9] Botez, *Românii despre ei înşişi.*

[10] This was the case, for instance, in 2002, with the fuel subsidies for mechanized agriculture, which were still being processed long after work had been completed.

[11] Arnăuţoiu Ion, "Arnăuţoii, Istoria unei familii," 2–3.

[12] Mihalache, "Cum să facem azi politică," 68–73.

[13] Caragiale, "Din primăvară pînă-n toamnă."

[14] Mungiu-Pippidi, *Politica după comunism*, 195–9.

[15] Iorga, *Supt trei regi*, 21.

[16] Mungiu-Pippidi, *Politica după comunism,* chs 4–5.

[17] Panel survey, autumn 2000, designed by the author and executed by CURS, available at www.curs.ro.

[18] Mungiu-Pippidi, *Politica după comunism*, 11.

[19] Huntington, *Political Order in Changing Societies*, 292.

CHAPTER 8

Between the Past and the Future

Seven million peasants, who barely reach subsistence levels, have no idea of the difference between the political right and left, and buy on average a single toothbrush in a lifetime pose a development problem not just for Romania, but for the European Union in general. It is hard to believe that by the twenty-first century the peasant problem in Romania had not only remained unsolved, it had actually been recreated close to its historical original, in spite of efforts to eliminate rural underdevelopment. In 2000, the Romanian village compared unfavorably with the city in most categories: access to services, communication with the rest of the world, human capital, and health statistics.[1] By then, the great myth of a healthy life in the country had given way to overwhelming disappointment. The poor quality of drinking water, the lack of domestic hot running water, and the difficult access to good-quality health care mean that the life expectancy of peasants is nearly two years less than that of city dwellers, and that peasants are more prone to diseases. If we take rural underdevelopment in Romania to be the last century's main unsolved problem for the country, then we must question to what extent there ever was a political will to solve it. The issue was certainly on everybody's agenda, from the liberals to the conservatives, and from the agrarianists to the communists and postcommunists. But it is its place in the hierarchy of issues on the political agenda which explains its resilience. It appears quite clearly now that the priority for the communists, by far the strongest regime, was not to solve the development problem as much as to assume control of society. Consequently, the creation of a substantial proletariat—which Romania lacked entirely at the end of World War II—was the key issue. The one of rural development was subordinated to the purely political project of subjecting the peasantry and to the chiefly economic one of bringing agricultural productivity closer to Western levels, in order to increase its contribution to national income.

Communism was successful in bringing the peasantry to its knees politically. The material condition of the peasant homestead was irrelevant, given that the regime's priority was to destroy any economic role that it might have had. In the hands of unscrupulous politicians, political liberalization did not contribute to a real emancipation of the peasantry, either in 1918 or after 1989. By enfranchising the peasants, the historical Liberal Party believed that it was securing the rural vote in the first elections based on universal suffrage. The concomitant land reform was supposed to give peasants the economic autonomy that would make them enlightened citizens capable of meaningful political participation. It was impossible to avoid land reform, when over half a million peasants had fallen in World War I, the Bolshevik Revolution was in full swing just beyond the country's eastern border, and the Romanian populists were creating quite a stir with their inflammatory speeches. In addition, as Radu Rosetti, Constantin Stere, Nicolae Iorga, and Ion Mihalache correctly observed at the time, Romania did not merely have large estates, but real latifundia. Yet, it was not difficult to foresee that land distribution would fail to create a class of farmers and would have a negative impact on agricultural production. Two-thirds of Romanian peasants were illiterate, the reproduction rate per family was extremely high, and basic institutions such as agrarian credit banks and competent "input-level" suppliers of agricultural services were lacking. Farming tools were primitive, and the size of the plots, as well as the credit shortage, made it difficult to purchase agricultural machinery. The "historic problem" of the Romanian peasantry was being unanimously identified with the shortage of land, lost to "rapacious boyars,"[2] but this was a subjective perception which corresponded only partly to reality. It was not land shortage, but the compulsory labor (corvée) on the boyar's land which, added to labor on the peasants' subcontracted land, made it impossible for the peasants to earn enough to buy land. Also, even if more land would have been made available, there were simply too many peasants: Due to the absence of any significant industrialization in the nineteenth century Romania had an agricultural overpopulation by 1900. The first consequence of abolishing the corvée was that pay for agricultural labor plummeted. And still the inability to create employment opportunities outside agriculture for the rural population persisted. Andrew Janos was right to say, when examining the transition from the nineteenth to the twentieth centuries, that Romania's development was inferior to that of its neighbors even in that period of relative boom.[3] There was no industry to attract peasants into the city, and there was added political tension as the towns and cities continued to be populated largely by a non-Romanian population of Greek and Jewish immigrants, living uneasily with the ethnically Romanian population of the countryside. Romania's historical failure to create towns prior to the nineteenth century was reinforced in the twentieth and was largely responsible for the difficulties of the peasantry. The absence of an ethnic Romanian urban population deprived the country of a counterweight to the regime of large landowners, who can be regarded as Romania's earliest "predatory elite" in a historical model defined by Henri H. Stahl as "the Prussian way to agricultural capitalism."[4] Henry Roberts was right to see the first half of the nineteenth century in Romania as a time of lost opportunities. Later, the political expression of the landowning class, the Conservative Party, was equally slow in deciding to abolish the corvée. The reforms of Cuza and Kogălniceanu came too late and achieved too little, and subsequent reforms collided with the intractable issue of overpopulation.

Further attempts to solve this problem in the interwar period overlapped with the Great Depression and the crash of grain prices worldwide (within a single year, in 1929, the price fell from 70,000 lei a wagon to 20,000 lei). We cannot say if the project of political Peasant Party leader Virgil Madgearu to create peasant cooperatives on a voluntary basis might have worked. What is certain is that twenty-five years after the agrarian reform of 1919–1921, the moratorium on farming debt, and countless lessons in agriculture delivered in the countryside by itinerant teams, the paternalist regime had still not solved the issue of rural overpopulation, although it had managed to strengthen individual peasant property. The communist rival project, initiated after 1945, undid these small achievements and brought the peasantry into a state of total economic and political subjection to a second "predatory elite." In doing so, it seriously affected its lifestyle and its identity as a social class. The communist state—as Stahl used to observe in private conversation, the only forum for such views during communism—had replaced the erstwhile feudal lord.[5] Ultimately, the postcommunist regime, with greater resources than previous regimes, failed even more miserably, reducing the peasantry to abjection.

This latter failure deserves closer scrutiny. Leaving aside our two villages, in this final chapter we shall consider agrarian policies as a whole from 1990 to the present day. At the same time, we shall look at the equally important aspect of the system of incentives and constraints which led to the choice of these particular policies rather than others.

Agrarian policy 1990–1996

After the 1989 Revolution the socialist legacy in agriculture proved difficult to divide among its natural heirs. In early 1990, agricultural land, still the chief Romanian resource (over sixty-five per cent of the country's land is agricultural, of which eighty per cent is arable), was divided between two categories of inheritor. The first included over 5,000 agricultural cooperatives, created out of individual plots voluntarily surrendered or taken by force after the last war. In the second category were the state farms, created from what had remained of the bigger estates, some individual peasant plots, and new terrains regained after the clean-up of morasses in areas such as the Great Lake of Brăila. Law 18/1991 protected, as we have seen, the state farms which had gained from the incorporation of private lands. Former owners were given shares as compensation for their lost land. But the law had major flaws, which undermined its successful application: As shown earlier, by 1995 not even half of all title deeds had been issued, and by 2000 the final target was still far from being met. It was not just those flaws which delayed the law's application, but also the passive opposition of the bureaucracy, especially at the prefectures. The rural councils committed mistakes and abuses in returning land, and the prefectures—which are accountable directly to central government—delayed issuing the title deeds for years after the councils had already solved matters satisfactorily. Even leaving aside private lawsuits over property arising from legal ambiguity, gaps in the archives and the poor performance of land restitution commissions, the complaints against the county commissions accountable to the prefectures amounted to over 200,000 (Table 14). The delays, combined with the statutory ban on the sale of land to prevent people from legally selling land until 1998, put off the creation of a land market.

Table 14. Complaints against decisions by the county commissions created under Law 18

Year	First Instance Tribunal		Court of Appeal	
	Lodged	Settled	Lodged	Settled
1991	40,413	17,779		
1992	51,208	54,977		
1993	35,037	34,912	11,453	3,545
1994	27,819	28,244	10,232	15,460
1995	23,404	24,512	7,968	8,049
1996	14,947	17,971	5,705	6,143
1997	10,946	13,119	4,219	4,851
1998	7,075	10,215	3,217	3,337
1999 1st term	1,814	2,040	703	744
Total	212,613	208,269	43,497	42,129

Source: Composite data from the archives of the Justice Ministry.

Besides the land itself, the entire inventory of farming equipment was gradually transferred into private hands. Ten years after the Revolution of 1989 around half of the pool of tractors was in private ownership, but by that time the pool dwindled to eighty per cent of what it had been at the end of the communist period, notwithstanding replacements and new acquisitions. The same decline was noticeable in irrigation equipment.

Exempt from privatization were all the state support companies, from the input level (producers of seeds, fertilizer, and farming machinery for instance) to the output level (including wholesale buyers of grain). Firms such as Romcereal—the main integrator firm in agriculture—became monopolies or, as the case might be, monopsonistic firms. The state, therefore, retained its control over important instruments which it could deploy at will to intervene in agriculture. And the state decided on such an intervention after the drought of 1991, in order to prove that the agricultural market was malfunctioning on its own and that it was in desperate need of a helping hand from the state.[6] Consequently, instead of liberalization of the market, the private agricultural sector was placed in the "straitjacket of being practically state institutions acting as monopsonistic firms in the acquisition of inputs and the collection of outputs alike."[7] The ideal template for this intervention was created by means of an obscure law called the "Law on state support for agricultural producers" (Law 83/1993). In conjunction with Government Ordinance 353/1995, which established prohibitive taxes on agricultural imports, Law 83 offered automatic and comprehensive assistance to producers, from free allocations of fertilizer and soil enhancers to subsidized minimum guaranteed prices, subsidized interest rates on loans, and "output bonuses." Like most postcommunist laws, this one, too, was passed without an assessment of the requisite budgetary resources and, as a result, proved largely inapplicable. It was one of a set of populist laws promoted by the postcommunist legislatures which were then quietly improved upon through selective application by the executive. As one analyst observed, the law implied practically "that agriculture is entitled to a redistribution in its favor of a proportion of the resources generated by other sectors of the economy."[8]

The mix of controlled prices, taxes on imports and exports, and a centralized, planned approach to agriculture from 1990 to 1994 has led, as happens in planned economies, to unforeseen effects. A World Bank study of 1994 showed that the aggregate effect of government intervention in the agricultural sector in 1992 amounted to a net transfer of considerable[9] resources away from agriculture towards the rest of the economy. This was done chiefly through capping prices for agricultural produce—in order to protect consumers—which led to an artificial compression of farmers' incomes. Between 1990 and 1996, the Romanian transition was distinctly marked by interventionist laws with contradictory, paradoxical, or downright unwanted outcomes. At the same time, the state merely went through the motions in playing what should have been its crucial role in the recreation of property market. The combination of chaotic interventionism and political reserve in matters related to private property prevented until 1998 the emergence of a land market, a process which is still incomplete today, owing to delays in issuing property title deeds. Subsidies were wasted in bailing out inefficient and underperforming companies such as farms and agricultural firms, especially in the state sector. The main bank which was supposed to support agriculture, the Agricultural Bank, was practically bankrupt by 1996, having built grand marble-decorated branches throughout Romania and given loans to the underperforming agricultural firms of the political establishment's clientele.[10] In 1997, the Victor Ciorbea government had to channel cash once again from other sectors to cover the losses of the Agricultural Bank, in response to fears of an imminent collapse of the banking system—something which had already occurred in Bulgaria. The bank was later privatized in a low-valuation deal.

The combined effect of such transition policies was a twenty per cent contraction of agricultural output between 1989 and 1998. Yet revenues from agriculture were not too seriously affected. Productivity was low: By 1998 only half the fields had been treated with fertilizers and pesticide. However, many of the poor-quality Romanian-made products, such as tractors manufactured by Roman Brasov, which were hardly suitable for small plots, and countless other goods of doubtful value for money, were all stuffed down Romanian farmers' throats. The state not only intervened, but did so in an absurd manner, trying to salvage the autarchic economy left by communism, and not realizing that what it gave to some sectors it took from others, thereby undermining the functioning of the market everywhere. An Agency for Rural Development, projected under Law 18, was to be given pre-emptive rights to buy back land from those unwilling to farm it and sell it on. But its task in creating economically manageable plots using the methods of the planned, centralized economy was so outdated and unfeasible that the agency ultimately never materialized. Real life, with its informal sales of land, was already miles ahead by the time plans for the creation of the agency were scrapped. At the same time, the integrator firms had accumulated debt after debt, and the preferential loans given to state farms were clearly never going to be paid back.

The postcommunist governments committed, on a larger scale, the historic error already made by the interwar regimes, and defined by the most prominent of Romanian economists, Virgil Madgearu, as "chaotic, spasmodic state interventionism, in which anything goes." In 1940, the agrarian economist saw two possible choices for the country's economic future: "the path of pure liberal-capitalist economy or state-controlled, planned economy."[11] The centrist government which succeeded the postcommunists in 1996 obviously had only one of these choices left. The other had been experimented with for fifty years, with grave consequences for Romania's economy.

Agricultural politics after 1997

The post-1996 centrist government was based on a coalition of parties. Consequently, its policies with respect to the peasantry and agriculture reflected, sadly, the absence of a clear consensus among the parties. The National Peasant Party (NPP) was the inheritor of the historic peasantism of the party's founders, Ion Mihalache and Virgil Madgearu. They were in the mainstream tradition of Romanian populism, and therefore the small peasant landholders constituted their constituency, although this category had been totally destroyed by communism. They sought to recreate it. In 1990 and 1991, the NPP—notwithstanding the fact that it was a party composed of gulag survivors—was highly vocal in deploring the farmers' associations scheme promoted by Ion Iliescu and in encouraging the peasants to take over their former lands completely and individually. The party's post-1990 program was guided by the illusion that the restoration of the small peasant holding would recreate the agrarians' electoral base of 1946, when most peasants in Nucşoara and Scorniceşti, alongside large sections of the Romanian peasantry, voted for the NPP.

The second party in the coalition with an interest in agriculture was Petre Roman's Democratic Party (DP), which split in 1991 from Ion Iliescu's postcommunist party. The Democratic Party had many state economy managers among its membership, and many more joined after the party won the elections. Triţă Făniţă, a representative of a powerful group of state farm directors, emerged as the most influential DP member and a defender of the old social order. He promoted policies which differed considerably from those of the National Peasant Party and, although publicly disowned, he was never really silenced by his DP leaders.

Therefore, after 1997, Romania had a succession of governments which were politically weak because they were based on undisciplined parliamentary coalitions, and administratively weak because they had inadequate resources to implement policies. What these governments did was attempt to put into practice programs designed by international aid agencies, such as the World Bank's ASAL and rural development programs, and programs proposed by the International Monetary Fund and the European Union. Agricultural and development policies were tackled separately, but, at least, it was the first time one could speak of governments having development policies at all. Pre-eminent in this respect were the European Union's pre-accession programs designed to prepare Romania for spending adequately the structural funds for rural development after accession. Yet, because of governmental instability and weak capacity on the Romanian side, at the end of the mandate of the only government not to have roots in the communist past rural development aid was barely emerging from the research phase. Research was, of course, very necessary: It divided the country into agro regions (areas with a similar farming profile) and "statistical" NUTS regions after the European model (clusters according to development indicators). In particular, research showed that differences between these areas were small and that the country was fragmented in islets of uneven development which overlapped badly or hardly at all with the traditional, historic regions.[12]

Policies meant to strengthen efficiency in agriculture and consolidate small farms were thus designed to correct the policy failures of the pre-1996 period. The second land law, in particular, finally authorized the sale of lands and buried the unborn Agency for Rural Development. Up to 1997, agriculture had been managed by agronomists, in line with the entrenched idea—inherited from communism—that only trained agricultural technicians can be good managers in agriculture. One of Romania's few economic analysts of agriculture complained bitterly of the absence of economists from the picture: "Debates among 'initiates' broadcast in the public media focus *ad nauseam* on specific aspects such as the 'optimal' amount of fertilizer per hectare, the 'optimal' period for applying herbicides, the 'optimal' distance between rows of seeds. The Ministry for Agriculture and Food, the institution chiefly responsible for formulating agricultural policies, employs an overwhelming number of agronomists."[13]

After 1997, a better policy re-established the role of economists in this process, only such experts were almost entirely lacking. In 2000, Editura Economică, the Bucharest-based publisher of economic literature, published a manual for the Academy of Business and Commercial Studies (where Madgearu was once professor) in which Romanian agriculture is discussed in the chapters "Labor," "Capital," "Land," and "Indicators." Significantly, the only chapter which, as the title "Agricultural Policies" suggests, should have treated Romania's agricultural policies—the book's topic—merely outlines the European Union's common agricultural policy. The general impression left by the volume is that the state of Romania's agriculture, highlighted in the tabled figures, is somehow beyond human intervention. This also explains why one of the major reforms of the post-1997 period, the transformation of the Ministry for Agriculture into an agency for agricultural services, was successful only in small part. The trained experts to staff such services were lacking, pure and simple.

The greatest failure, however, was the government's inability to privatize the state farms. The draft laws on state farms, as well as on the restitution of forest land, were the object of fierce battles in the parliamentary commissions, presided over by Triță Fănță. As the governing coalition did not have a majority without the Democratic Party, and as the DP de facto joined forces with the opposition postcommunists on this particular point, this sealed the fate of both state farms and the forests, controlled by the state monopoly, the Forestry Authority. The two laws were delayed until 2000, which led to accrued debts and arrears for the state farms. The delays also gave an opportunity for the postcommunists, who were returned to power in the autumn of that year, to intervene once more, by restricting the applicability, and therefore also the outcomes, of this legislation. Ultimately, these outcomes largely offset the positive effect of the reforms. As far as private farmers were concerned, two measures had immediate positive results: one was lifting the ban on land sales and the other was a rechanneling of subsidies towards the private sector by means of the "voucher" scheme. Agricultural policies in the period studied generally suffered from two structural flaws in the government's overall economic policy: improvisation and a lack of confidence in the processes of the market, which led to many half-based and inconclusive actions.[14]

A vicious circle renewed

Any policy regarding the peasants is, openly or not, grounded in some particular ideology. Even the most evidence-based, utilitarian-driven policy cannot escape a degree of ideological touch, as it has to draw on some vision on collective action. Yet, as Geertz observed, even the most partial or the most immoral of ideologies has something to teach us about their object, something which otherwise might remain difficult to grasp.[15] In the same way, a metaphor cannot make any claims to precision, but it may shed light on areas which might have remained closed to science but are worth knowing about. For example, when reading what the communists had to say about the peasants, in the available minutes of genuine discussions rather than in stiff, official documents, one finds that their criticisms are not far removed from the bitter reflections of interwar authors such as Dumitru Drăghicescu and Constantin Rădulescu-Motru. The first First Secretary of the Communist Party, Gheorghe Gheorghiu-Dej, for instance, said: "Our peasants are happy living from hand to mouth, but we have to make them work because work will improve their living standard and their political standard, while also contributing to the country's economic growth."[16] Elsewhere, he said: "If his child dies, the peasant will be mourning less than if his cattle die. There are places in this country where cattle enjoy better living conditions than the peasants. [...] The man walks ahead with his walking stick, like a petty master, while his wife walks five paces behind, carrying a heavy bag and a child in her arms."[17]

Those who worked to raise the development level of rural Romania in the interwar period—the paternalist stage in dealing with Romanian agriculture and peasants—were equally perplexed, although more polite about it than Dej, seeing the peasant's resistance to attempts at changing his lifestyle and working habits. The founder of Romanian agronomy, Gheorghe Ionescu-Siseşti, deplored the "routine" of the peasant household, while his student Nicolae Cornăţeanu (his future successor as Minister of Agriculture) reflected sadly in the *Enciclopedia României*: "The main role of peasant agriculture is to take full advantage of the labor resources of the family. Hence the peasant landholder needs to privilege labor-intensive cultivation. Our present peasant agriculture is based on a system of cultivation which does not correspond to a modern agrarian structure [...] The peasant [...] does not copy any of the farming techniques routinely practised on the large estates. Therefore, modernizing peasant farming raises major social and psychological problems."[18]

The ideologies about the Romanian countryside can be grouped into two main categories and a variety of subcategories. The first one is a modernizing ideology, which covers a range of subcategories including liberalism and communism. It regards the peasants as emanating from a particular stage in social development, one which is destined to disappear as technology advances and productivity increases. In this view, the peasant and his household are seen as obstacles to progress and must be either transformed by turning peasants into farmers, for instance, or must disappear altogether through urbanization, cooperativization, etc. Champions of such views, from modernizing agronomists such as Ionescu-Sisești to brutal reformers such as Dej, believe that peasants are essentially passive and therefore incapable of political participation. In any case, modernizers placed the emphasis on industry. For them, agriculture is secondary, and their tariff policies reflected this ideology.

The second ideology is the populist one—with many subvarieties in Romania—according to which the peasantry is essentially capable of political action and even of self-government, provided it has a guaranteed socioeconomic base. In addition, populists in all subcategories believe that the peasantry is an important part of the Romanian identity and will remain so in the future. Peasantism is only one ideological subvariant of Romanian populism, as Ghiță Ionescu pointed out.[19] There are pragmatic populists such as Ion Mihalache and the agrarianists in general, there are populists of a mystical orientation, such as Mircea Vulcănescu and Horia Bernea, but the views outlined above are characteristic of the populist vision as a whole, irrespective of differences among its subvarieties. On the few occasions when they found themselves in government, the populists relaxed tariffs and attempted to eliminate the protectionism which favored industry at the expense of agriculture. But the period in which they could have had impact on policies coincided with the worst ever global economic crisis. The meager outcomes of such attempts made Virgil Madgearu—who paid with his life for his globalist economic views[20]—come to the bitter conclusion that free market was utopian and Romanian economy could be saved only by total state control.

In spite of wide differences in political freedoms from one regime to another, all regimes were net contributors to peasant dependency. Therefore, all peasants, landless and landholders alike, have failed to come to enjoy genuine political autonomy and have remained dependent on a local elite with disproportionate political power. The sources of this dependency varied over time: First it was the illiteracy of peasants and the strength of traditional authority (before the war), then the lack of economic sustainability of their farms, then various forms of coercion, more or less subtle (before the agrarian reform of 1919, under communism and postcommunism). The traditional authority was largely destroyed with the prewar elites, and the only remaining mechanism for controlling the peasantry politically was a combination of coercion and patronage. The postcommunist period further refined this mechanism: Populist legislation signaling political support for the peasantry was accompanied by measures which pumped state subsidies into state agencies and farms, de facto channeling them towards the predatory elites which controlled such institutions. Given that in 1990 the peasant vote went unanimously to the former Communist Party secretary Ion Iliescu, the paternal figure identified with the state, such a development, of all the other developments in the transition period, was not inevitable. While Ion Iliescu may have looked like the guarantor of order who sent the Army back into the barracks at the Revolution, he certainly did not depend for support on the state farms and the managers who composed the predatory elites. Both Ion Iliescu and his party, as well as the Democratic Party until 1999, made an unambiguous political choice when they decided to give their support to the predatory elites. The state of dependency of the Romanian peasants at the end of the transition period was an outcome that resulted partly from policy failure, but partly from design. It came about as a result of a disastrous combination created to a large extent by Ion Iliescu's long reign over Romania's postcommunist politics.[21] Despite Iliescu's personal claims to honesty, matched by his rigorous and unostentatious lifestyle, he was the undisputed patron of such elites. His retirement in 2004 after three terms as president (one shorter, 1990–1992) opened for the first time the door to some genuine political competition in the countryside, as the elites had to reorient themselves and find a new patron.

In 2002, the young mayor of Scorniceşti was sitting in his last-generation office chair under a large portrait of Iliescu, with another framed one on his desk. Whose portrait would sit there next? As we called on peasants in their homes, our guide, a young civil servant from the mayor's office, was greeted with: "I haven't seen you since the campaign, when you came to tell me whom to vote for." In Scorniceşti, as well as in Domneşti, the predatory elite was already looking past Iliescu to Vadim Tudor and the Greater Romania Party. They share a large chunk of history.[22] In contrast, the National Peasant Party was led so badly that it was practically pulverized in the elections of 2000 and subsequently descended into almost total obscurity. The Liberal Party has got a powerful branch in Scorniceşti town, but, as an essentially capitalistic and urban party, it garners no votes in the neighboring villages. The Democratic Party, which broke with Iliescu in 1991, still enjoys the allegiance of rural elites, and their leader Traian Băsescu, a former ship's captain elected President of Romania in 2004, fits the image the peasants have of a leader quite nicely. By 2009, after losing most of his urban support, Băsescu was defending his office for a second term based mostly in the countryside.

European prospects

Romania was invited to join the European Union in December 1999 at the Helsinki summit. Accession negotiations started the next year and were the slowest among the twelve countries invited to join. Accession is a technical process, involving a regular review of the progress made by a candidate country in adopting the *acquis communautaire*, the total body of EU law accumulated so far. The *acquis* is adopted by the national parliaments in a first stage. Subsequently, it has to be transposed into each state's domestic legislation and applied. Of the new candidate states, Romania had the hardest time concluding negotiations and implementing the *acquis*, although, together with Bulgaria, it was anyway scheduled to accede three years later than the other postcommunist countries. Accession strained the already low capacity of the state and tested the limits of political will. However, since popular support for Europe, seen as a rescuer, particularly in material terms, has always been strong, no party, including Mr. Iliescu's, could afford to be perceived as failing at it. Policy distance between parties thus decreased considerably, as they all had to implement the same legislation and struggle with the substantial difficulty of having to catch up economically as a country where GDP was a quarter of the EU average by 2000. Some of the more serious difficulties lie in the harmonization of agriculture and rural development. Allowing these areas to sort themselves out is unlikely to prepare Romania for European accession: only consistent, coherent, and confident policies, based on cross-party consensus and EU support, can gradually solve the difficulties—both inherited and created—in the Romanian countryside, as well as their effects on the rest of the economy. The most serious challenges to Romania's European integration lie in the countryside.

Problems of Romanian agriculture in the context of EU accession

In 1989, the year when Nicolae Popa's statistical data were being further inflated in Bucharest, Romanian agriculture produced 33.6 tons of wheat per hectare. In the same year, another socialist country, Czechoslovakia, produced an average of 52.6 tons, compared with 61.5 in France and 68.4 in Germany. Today the yield in Romania is even less because of the structures of property and lack of adequate inputs. In other words, Romanian agriculture is still underperforming. Yet it is a very important part of the country's economy. Its average contribution to GDP during transition was around nineteen per cent, well above not just the EU average but also that of neighboring countries. The proportion of the workforce engaged in agriculture is also considerable: At thirty per cent it is, for instance, well above Poland's (thirteen per cent), a country which most closely matches Romania's economic profile in the region (Table 15). In spite of the steady development of the services sector over the last decade, the contraction of the state sector—suffocated by debt arrears—and the pace of growth in the private sector mean that cities have not managed to create sufficient job opportunities for the rural workforce. The urban private sector struggled primarily to absorb the workers made redundant in the state sector. The countryside has been left to itself, and has to be reinvented economically from within. The entrepreneurs and services likely to create growth will have to operate from the countryside itself. Development on the basis of migration into cities has exhausted its possibilities. Only upgrading and modernizing the countryside is likely to solve the problems of rural underdevelopment and farming in Romania.

Table 15. The comparative position of Romanian agriculture in the region

	% of workforce	GDP per capita in Purchasing agriculture[1] Power Standard (PPS) (EU 27 = 100)[2]	% rural residents[3]
Bulgaria	7.42	40.2	29.74
Czech Republic	3.14	80.1	26.5
Estonia	3.85	68.2	30.86
Latvia	8.95	55.8	32.12
Lithuania	9.27	61.1	33.44
Hungary	4.56	62.8	33.3
Poland	13.38	57.6	37.76
Romania	30.12	45.8	46.12
Slovenia	9.16	90.7	48.82
Slovakia	3.63	71.9	43.68

[1] Data for 2009, Source: Eurostat
[2] Data for 2008, Source: Eurostat
[3] Data for 2006, Source: UNESCO

Trade development with the EU, as well as with members of the Central European Free Trade Agreement (CEFTA) which replaced the Council for Mutual Economic Assistance (Comecon) of socialist times, has highlighted the low competitiveness of Romania's agriculture and food industry. Romania imports far more than it exports to CEFTA countries. Only in Hungary's case could it be claimed that the higher competitiveness of Hungarian agriculture is due to higher state subsidies. Most Romanian products are below the standards required by the EU, which is a problem even for exports outside the

EU. Achieving higher competitiveness will depend on the continuation of structural reform and on the creation of a functional agrarian market which works on the basis of competition rather than of preferential loans.[23] Slogans heard in Bucharest about the massive subsidies to agriculture in EU countries ignore one essential detail, namely that EU countries can afford to subsidize their agriculture, given their economic profile, which differs so profoundly from Romania's. In the EU, non-agricultural sectors account for 90–97 per cent of the total workforce, hence the relatively low cost of subsidizing the remaining 3–10 per cent. In Romania, this ratio is 2:1, and subsidizing agriculture would be like two people having to pay to maintain a third. Another fundamental difference is that the two are not doing that well in the first place. In the European Union, a four-member household "transfers" 1,000 USD annually to agriculture, which would be impossible in Romania, where total annual income per household is 1,500 USD. The same could be said of consumer subsidies. In Romania, the consumer spends 58 per cent of household income on food, and so cannot afford what the Hungarians—who spend only 24 per cent on food, or EU citizens with only 17 per cent—can easily afford.[24] In brief, the countryside cannot be subsidized by the city.

The dependency model and European integration

In a society which is only partly modernized, dependency, which prompts peasants to vote for those who oppress them, is part of a more general pattern of dependency and particularism afflicting the entire society, not just the countryside. This pattern of a rural culture extended to society as a whole poses serious problems in countries aspiring to European integration. The model described in detail in the present study survived the coercive communist system and prevents the creation of a genuine rule of law. Romanian postcommunist elites considered EU accession a good bargain, and converted themselves into Europeans. But how deep has this conversion been? Barely a few weeks after accession in 2007, the parliament tried to fire the Justice Minister and the anticorruption prosecutor credited for convincing Brussels that the country is fit to join. The political history of Romania from 2004 to the present, as reflected in the Freedom House Nations in Transit reports, is a story of an unfinished struggle between predatory elites and new EU accession-driven institutions.

European integration is correctly perceived in Eastern Europe as a last phase of modernization, emphasizing an altogether different type of culture from communism, based on an impersonal rule of law and an impartial government. Can a country such as Romania, which has incompletely modernized in terms of political and administrative culture, achieve such a substantial change in the space of only a few years? Is European integration an incentive strong enough to change such predatory behaviors and make them obsolete? Alternatively, can Romania gain access to the European Union without changing? Corruption has often been singled out by the European Commission as Romania's main problem. But the term "corruption" (the use of public office for personal gain) applied to Romania is misleading, as the phenomenon encountered is a systemic arrangement, not some individual infringement of the norm of bureaucratic impersonality and fairness. If we consider good governance the process by which a government distributes public goods equally and fairly to all citizens, ranging from their legal treatment to health and education, the opposite results: This is particularism or systemic corruption. In the former type of society, individuals expect equal treatment from the state. In the latter, their treatment depends on their status or position in society. Peasants do not even expect to be treated fairly by the state, what they expect is treatment similar to that received by everybody with the same status. Favors—i.e. discriminatory treatment—are not always offered for money, they can also be a means of creating obligation, a reward, a recognition of superior status, or a means of establishing superior status. The distribution or withdrawal of favors is part of the arsenal of traditional authority. Every group or society steeped in traditional authority—i.e. every premodern society—will have its "godfathers." When representatives of the modern bureaucracy behave like "godfathers," they do not necessarily expect monetary profit or influence-peddling. Quite often, they wish only to affirm or enhance their own status. A culture of privilege reigns in societies based on particularism, making unequal treatment the accepted norm in society. Individuals struggle to belong to the privileged group rather than change the rules of the game. There is a widespread infringement of the norms of impersonality and fairness, but also, depending on the openness and mobility, equilibriums are easily reached and are stable, making these regimes difficult to change. We know that societies undergoing transition have competing normative regimes which strive to become dominant. We know very little, however, of how universal government historically succeeds over the norm of particularism and what determines this development. A survey of developing countries returns rather a pessimistic answer to this question.

Being an EU member or not, Romania will face tremendous difficulty in changing the model described in this study, the model according to which, in the words of Ion Mihalache, "politics [...] is a matter of seeing to personal interests at the expense of public interests." The capture phenomenon includes law enforcement agencies, the judiciary, and even the private media. The incidence of such behavior across the political spectrum, as well as the frequent defections from one party to another of politicians and members of the managerial class, suggest that there are no fundamental differences in political culture among parliamentary parties. Even the more enlightened politicians in all parties appear unable to propose and support a change of system. In Romania and other East European countries, an elite alternative to the communist one was destroyed in the Stalinist era. But in Romania (and Albania) the process continued well into the late years of the communist regime, at a time when elsewhere in Eastern Europe timid oppositions were burgeoning under perestroika. In 2000, Romania lacked significant political alternatives to the predatory elite, and the foundations of collective action, largely dismantled by communism, were far from resurfacing. It takes time and political will to bridge the distance from decent people and far-sighted politicians to organizations really capable of changing the system, and from individual or group interest to the public interest. The champions of such changes are lacking in Romania. Intellectuals fixated on historical identities and on the ideas of Eliade, Noica, and Cioran, and a Church with delusions of grandeur, whose priests dream of state salaries, will not become the heralds of change. The only likely partisans of a paradigm shift are a handful of embattled entrepreneurs in a private sector which does not benefit from sufficient state support. They are carrying the Romanian economy on their backs, but, because of a dysfunctional system, they are unable to produce enough to narrow a huge historic gap. They are the only likely promoters of an open society against a status-based society.

It is to be expected that Romania, like Greece before it, will develop gradually, but without its fundamental institutional arrangements changing radically. The country will remain particularistic. Its rural problem is unlikely to be solved by policy means as much as by natural means. The dependent peasantry from subsistence farming areas is an aged population, prone to biological extinction in the next ten years. Migration to the informal labor markets of Italy and Spain has removed from the country about two million Romanians since 2005 (when Schengen visa requirements were lifted): Most of them work in agriculture and construction. The old village will die, being gradually replaced by the new town: Nucşoara will fade away and Scorniceşti will transform.

Notes

[1] Sandu, *Spațiul social*, 144–9.
[2] Stahl, *Satele devălmașe*, vol. 4.
[3] Janos, "Modernization and Decay in Historical Perspective," 96.
[4] Stahl, *Satele devălmașe*, vol. 3, 375.
[5] Interview by the author with anthropologist Paul Stahl, son of Henri H. Stahl, April 2002.
[6] See Teşliuc, "Agricultural Policies: Achievements and Challenges," section 3.1.2.
[7] Ibid.
[8] See Negrescu, *Politica agricolă*.
[9] Ibid.
[10] The most celebrated of these was the "Agricola" holding in Bacău, which represented the interests of top postcommunist politicians such as Dumitru Sachelarie and Viorel Hrebenciuc.
[11] Madgearu, *Evoluția economiei*, 400.
[12] See Sandu, *Spațiul social*.
[13] Private interview with Dragoş Negrescu by the author in 2001. Negrescu worked for the European Commission Delegation in Bucharest at the time.
[14] Teşliuc, "Agricultural Policies: Achievements and Challenges," section 3.2.2.
[15] Geertz, *Interpretation of Cultures*.
[16] Catanus and Roske, *Colectivizarea agriculturii*, 154.
[17] Ibid., 155.
[18] Entry for "Politica agrară," in Gusti, *Enciclopedia României*, vol. 3, 154.
[19] In Ionescu and Gellner, *Populism*.
[20] He was assassinated by the Iron Guard in November 1940.
[21] Ion Iliescu won presidential mandates between 1990 and 1996 and from 2000 to 2004.
[22] See Mungiu-Pippidi, *Early Warning Report Romania*.
[23] For a very good action plan, see Teşliuc, "Agricultural Policies: Achievements and Challenges," 32–40.
[24] Both Negrescu and Teşliuc reach similar conclusions on this subject.

References

Arnăuţoiu Ion, E. "Arnăuţoii, Istoria unei familii din Nucşoara povestită de Elena Arnăuţoiu Ion." *Supplement to Rosturi* 34 (April 1993).

Bărbulescu, V. *Consiliul unic agroindustrial de stat şi cooperatist, formă superioară de organizare şi conducere a activităţii economice în agricultură: bilanţ şi perspectivă la C.A.P. Scorniceşti, judeţul Olt*, 1980.

———. *Scorniceşti, cuvînt înainte*. Bucharest: Editura Sport-Turism, 1988.

Blaga, L. *Trilogia cunoaşterii*. Bucharest: Fundaţia Regală pentru Literatură şi Artă, 1943.

———. "Permanenţa preistorică." *Saeculum* (Sept.–Oct./1943).

———. *Trilogia culturii*. Bucharest: Editura pentru Literatură Universală, 1969.

Botez, M. *Românii despre ei înşişi*. Bucharest: Litera, 1994.

Caragiale, I. L. "Din primăvară pînă-n toamnă," *Momente, schiţe, povestiri*. Bucharest: Editura Ion Creangă, 1977.

Cătănuş, D. and O. Roske. *Colectivizarea agriculturii in Romania. Dimensiunea Politică*. Bucharest: Academia Română. Institutul Naţional pentru Studiul Totalitarismului, 2000.

Clark, C. *The Conditions of Economic Progress*. London: St. Martin's, 1980.

Deletant, D. "Studiu Introductiv," in Oprea, M., ed. *Banalitatea răului: o istorie a Securităţii în documente. 1949–1989*. Iaşi: Polirom, 2002.

Douglas, M. *Risk and Blame. Essays in Cultural Theory*. London: Routledge, 1992.

Drăghicescu, D. *Din psihologia poporului român*. 1907, 2nd edition, Bucharest: Editura Albatros, 1995.

European Commission, World Bank, Ministry of Agriculture and Food. *Farm Survey.* Bucharest, 1996

———. *Private Agriculture in Romania. Farm Survey.* Bucharest, project financed by the PHARE programme of the European Commission. 1998. *Rural Development in Romania.* Ministry of Agriculture and Food, Bucharest.

Fitzpatrick, S. *Stalin's Peasants: Resistance and Survival in the Russian Village after Collectivization.* Oxford: Oxford University Press, 1996.

Foster, G. M. "What is Folk Culture?" *American Anthropologist* 55, no. 2 (1953): 159–73.

———. *Traditional Cultures and the Impact of Technological Change.* New York: Harper and Brothers, 1962.

———. "Peasant Society and the Image of Limited Good." *American Anthropologist* 67, no. 2 (1965).

———. et al. *Peasant Society: A Reader.* Boston: Little Brown, 1967.

Geertz, C. *Peddlers and Princes: Social Development and Economic Change in Two Indonesian Towns.* Chicago: Chicago University Press, 1963.

———. *The Interpretation of Cultures.* New York: Basic Books, 1973.

Gordy, Eric D. *The Culture of Power in Serbia: Nationalism and the Destruction of Alternatives.* University Park, PA: Pennsylvania State University Press, 1999.

Gusti, D., ed. *Enciclopedia României.* Volumul III. Bucharest: Imprimeria Națională, 1939.

Hirschhausen, B. von *Les nouvelles campagnes roumaines. Paradoxes d'un "retour" paysan.* Paris: Belin, 1997.

Humphrey, Caroline. *The Unmaking of Soviet Life: Everyday Economies after Socialism.* Ithaca: Cornell University Press, 2002.

Huntington, S. *Political Order in Changing Societies.* New Haven: Yale University Press, 1956.

Iancu, Gh., Virgiliu Țărău, and Ottmar Trașcă. *Colectivizarea agriculturii în România. Aspecte legislative.* Cluj Napoca: Presa Universitară Clujeană, 2000.

Ionescu, G. and E. Gellner, eds. *Populism: Its Meaning and Characteristics.* London: Weidenfeld and Nicholson, 1974.

Iorga, N. *Supt trei regi.* Bucharest: PRO, 1999.

Janos, A. C. "Modernization and Decay in Historical Perspective: The Case of Romania," in Jowitt, K., ed. *Social Change in Romania: 1860–1940.* Berkeley: University of California, Institute for International Studies, 1993, 72–116.

Jowitt, K. *New World Disorder or the Leninist Extinction.* Berkeley: University of California Press, 1992.

———. *Social Change in Romania: 1860–1940.* Berkeley: University of California, Institute for International Studies, 1993.

Kideckel, D. *The Solitude of Collectivism: Romanian Villagers to the Revolution and Beyond.* Ithaca and London: Cornell University Press, 1993.

Kroeber, A. L. *Anthropology: Race, Language, Culture, Psychology, Prehistory.* New York: Harcourt, Brace, 1948.

Kurtz, M. Jr and A. Barnes. "The Political Foundations of Post-Communist Regimes: Marketization, Agrarian Legacies, or International Influences." *Comparative Political Studies* 35, no. 5 (2002): 524–53.

Leonard, P. and K. Deema. *Post-Socialist Peasant?* London: Palgrave, 2002.

Lewin, M. *Russian Peasants and Soviet Power: A Study of Collectivization.* Evanston, IL: Northwestern University Press, 1968.

Madgearu, V. *Evoluția economiei românești după războiul mondial.* Bucharest: Independența Economică, 1940.

Mihalache, I. "Cum să facem azi politică," reproduced in *Ce politică să facem.* Bucharest: Litera, 1995, 68–73.

Mitranyi, D. *The Land and the Peasant in Rumania: The War and Agrarian Reform, 1917–1921.* London: Oxford University Press, 1930.

———. *Marx Against the Peasant: A Study in Social Dogmatism.* London: George Weidenfeld and Nicholson, 1951.

Moore, B. Jr. *Social Origins of Dictatorship and Democracy: Lord and Peasant in the Making of the Modern World.* Boston: Beacon, 1966.

Moore, B. Jr. *Injustice: The Social Bases of Obedience and Revolt.* White Plains: M. E. Sharpe, 1978.

Mousnier, R. *Les Hiérarchies sociales—de 1450 à nos jours*. Paris: PUF, 1969.

Mungiu, A. "Intellectuals as Political Actors in Eastern Europe: The Romanian Case." *East European Politics and Societies* 10, no. 2 (1996): 333–64.

Mungiu-Pippidi, A. *Politica după comunism*. Bucharest: Humanitas, 2002, ch. 5.

———. *Secera şi buldozerul.* Iaşi: Polirom, 2002.

———. *Early Warning Report Romania*. UNDP, no. 5/2002, Political Section.

———. *Nations in Transit: Romania 2003–2008*. Washington: Freedom House, http://www.freedomhouse.org/template.cfm?page=47&nit=464&year=2008.

———., D. Mândruța, and S. Ioniță. "In the Shadows: Survival Strategies of the Unemployed in the Transition," Working Paper, Vienna Institute for Human Studies, SOCO, no. 80 (2000).

Negrescu, D. *Politica agricolă a României în perioada de tranziție*. Bucharest: Centrul de Studii Politice şi Analiză Comparativă, 1995.

O'Donnell, G. "State and Alliances in Argentina, 1956–1976." *Journal of Development Studies* 15, no. 1 (1978).

———. "Illusions about Consolidation." *Journal of Democracy* 7, no. 2 (1996): 34–51.

Ogoranu, I. G. *Brazii se frâng, dar nu se îndoiesc*. Timisoara: Editura Marineasa, 1995.

Oprea, M., ed. *Banalitatea răului: o istorie a Securitătii în documente. 1949–1989*. Iaşi: Polirom, 2002.

Petrescu, General Gh. Fl. *Monografia județului Olt*. N.p., n.d.

Palairet, Michael. *The Balkan Economies c. 1800–1914: Evolution without Development*. Cambridge: Cambridge University Press, 2003.

Presidential Commission for the Analysis of Communist Dictatorship in Romania, *Final Report*. Bucharest, 2006, available at http://www.presidency.ro/static/ordine/RAPORT_FINAL_CPADCR.pdf.

Rădulescu-Motru, C. *Etnicul Românesc şi naționalismul*. Bucharest: Albatros, 1997.

Redfield, R. *Tepoztlan: A Mexican Village. A Study of Folk Life*. Chicago: University of Chicago Press, 1930.

———. *The Primitive World and Its Transformations*. Ithaca: Cornell University Press, 1953.

———. *The Little Community*. Chicago: University of Chicago Press, 1955.

———. *Peasant Society and Culture*. Chicago: University of Chicago Press, 1956.

Report of the Prime Minister's Control Department of Romania, July 1997.

Roberts, H. *Romania: The Politics of an Agrarian State*. New Haven: Yale University Press, 1951.

Rose, R., W. Misher, and C. Haerpfer. *Democracy and its Alternatives: Understanding Post-Communist Societies*. Baltimore: Johns Hopkins, 1998.

Rosetti, R. *Pământul, sătenii şi stăpânii în Moldova*. Bucharest: Atelierele grafice Socec & Co, 1907.

———. *Pentru ce s-au răsculat ţăranii*. Bucharest: Editura Eminescu, 1987.

Sandu, D. *Spaţiul social al tranziţiei*. Iaşi: Polirom, 1999.

Scott, J. C. *Weapons of the Weak: Everyday Forms of Peasant Resistance*. New Haven: Yale University Press, 1984.

Seton-Watson, H. *The Pattern of Communist Revolution: A Historical Analysis*. London: Methuen & Co, 1960.

Spălăţelu, I. *Scorniceşti. Vatră de istorie românească. Nicolae Ceauşescu fără drept la moarte*. Bucharest: Editura Viaţa Românească, 1999.

Stahl, H. H. *Satele devălmaşe* (vols. 1–4). Bucharest: Cartea Românească, 1998.

Stalin I. V. *Opere* (vol. 2). Bucharest: Editura PMR, 1951.

Stere, C. *Scrieri*. Bucharest: Editura Minerva, 1979.

Teşliuc, D. E. "Agricultural Policies: Achievements and Challenges," World Bank Poverty Reduction and Economic Management, 1999.

Verdery, K. *What Was Socialism and What Comes Next?* Princeton: Princeton University Press, 1996.

Voicu-Arnăuţoiu, I. R. *Luptătorii din munţi: Toma Arnăuţoiu, Grupul de la Nucşoara: documente ale anchetei, procesului, detenţiei*. Bucharest: Vremea, 1997.

Vulcănescu, M. *Prolegomene sociologice la satul românesc*. Bucharest: Eminescu, 1997.

Waldron, A. G. *Communist Neo-Traditionalism*. California: University of California Press, 1986.

Weber, M. *The Theory of Social and Economic Organization*. New York: Free Press, 1941.

———. *On Charisma and Institution Building*. Chicago: University of Chicago Press, 1968.

Wegren, S. *Land Reform in the Former Soviet Union and Eastern Europe*. London: Routledge, 1998.

———. "The Land Question in Ukraine and Russia," *The Treadgold Papers*. Seattle: University of Washington, 2002.

Werth, N. "Introduction. Le pouvoir sovietique et la paysannerie dans les rapports de la police politique (1930–1934)." *Bulletin de l'Institut d'Histoire du Temps Présent* 81–82 (2003).

Wolf, E. R. *Peasants*. Englewood Cliffs, N.J.: Prentice-Hall, 1966.

Appendices

1. Survey sources

The present study cites a number of opinion polls:

- County-level representative poll, October 2000, CURS (Romanian Centre for Urban and Regional Sociology) – CURS 1. The size of the sample was 37,474; for Olt county n=851 and for Argeş county n=848.
- National representative poll, November 2000, CURS and Romanian Academic Society – CURS 2.
- National representative poll conducted by Metromedia in May 2001 for the Soros Foundation (BOP Programme) – Metromedia.

Characteristic	National	Argeş	Olt
Education (years) average	4.13 (1.50)	4.39 (1.46)	4.24 (1.46)
Average age	46.34 (17.2)	44.18 (15.96)	45.32 (16.43)
Sex ratio	49:51	47:53	53.8:43.2
Rural/urban pop. ratio	53:47	48:52	60:40

Source: CURS 1

1.1 JA comparison of the regions studied with the national sample
Figures in brackets are standard deviations.

1.2 Comparative living standards: urban – rural – regional

Rural-urban comparative indices of ownership

Percentage owning	National	Urban	Rural	Argeş	Olt
Land	45.5	19.5	74.4	49.1	58.8
Agricultural machinery	3.0	1.0	5.2	1.5	3.8
Black-and-white TV	41.9	30.1	55.0	41.0	49.1
Color TV	64.5	82.6	44.3	70.9	60.4
Video	15.8	22.6	8.2	15.3	14.7
Aerial	48.2	25.7	73.3	46.2	63.9
Cable TV	45.0	77.5	8.6	48.0	31.1
Satellite dish	4.5	2.8	6.5	3.2	4.5
Several homes	5.8	6.7	4.8	5.2	7.8
Car	30.7	39.9	20.4	41.2	29.5
Telephone	49.4	73.8	22.1	37.1	28.6
Refrigerator	80.9	93.2	67.2	84.4	83.9
Freezer/freezer compartment	27.7	37.0	17.3	33.4	24.1
Automatic washing machine	22.1	34.0	8.8	20.0	14.7

Source: CURS 2

2. Nucşoara and Scorniceşti – Demographic and electoral village data

SCORNICEŞTI

Total population: 12,787
Urban: 4,794
Rural: 7,993

Villages:
1) Scorniceşti
2) Piscani
3) Baltati
4) Mărgineni-Slobozia
5) Mihăileşti-Popeşti
6) Constantineşti
7) Mogoşeşti
8) Negreni
9) Bircii
10) Chiteasca
11) Teiuş
12) Rusciori
13) Jitaru
14) Suica

NUCŞOARA

Total population: *c.* 1,800

Villages:
1) Nucşoara
2) Gruiu
3) Sboghiţeşti
4) Slatina

Index